Discrete Calculus By Analogy

Author: F. A. Izadi Coauthors: N. Aliev, G. Bagirov

December 3, 2009

Contents

Foreword

In this nice book it is dealt with Discrete Calculus, in particular with discrete derivative and integral and several models of discrete differential equations with different kinds of "Cauchy-type" problems.

The exposition is very clear and detailed, in particular in the calculus of explicit solutions. Furthermore the book contains some topics, specially in the appendix, which can be studied both by researchers and by undergraduate and/or Ph. D. students, in order to be introduced in problems of several branches of Mathematics, for example numerical methods to solve differential equations of Mathematical Physics and discretization of continuous models, which makes more accessible the study of phenomena related with them by means of the computer.

At the beginning of this self-contained book there is a precious and detailed introduction of remarkable didactic value, which is accessible even to students, who want to start to investigate the Discrete Calculus and related topics even without having any preliminary background, and in which some fundamental tools, like complex numbers, linear systems even of infinite equations, determinants, Cramer rule, and so on, are illustrated.

This book can be used also as a primer by researchers who want in a further step to study several applications, in particular to ordinary and partial differential equations, discrete (and usual) Fourier transform, eigenvalue problems, operators and spectral theory, Numerical Analysis, difference equations, finite and divided differences and several types of derivatives even in the "continuous" setting, like for example Peano and Riemann symmetric derivatives.

Another topic on which this book can give an inspiration, is a comparison between the functions, derivatives and differential equations here presented and the corresponding classical ones.

Antonio Boccuto,

Perugia-Italy

"Pythagoras theorem is the analog of the Parseval equality on a finite dimen-sional vector space." Statement made by a mathematician.

Preface

This little book has been written to present a concise and self-contained account of some topics in *"difference calculus"*. But for some reasons which will become much more clear at the end of this preface, we choose the title to be *"discrete calculus by analogy"*.

If the content is not so sophisticated, it is designed, as is clear from the above expression, to investigate some patterns of mathematical facts in discrete cases. These patterns and the methods of their investigation are not only interesting concepts in their own rights, but they can also be used to motivate the investigation of continuous analogs, from which one can search, compare, unify, and ultimately discover general facts.

The topics covered here usually arise in many branches of science and technology, especially in discrete mathematics, numerical analysis, statistics and probability theory as well as in electrical engineering, but our viewpoint here is that these topics belong to a much more general realm of mathematics; namely calculus and differential equations because of the remarkable analogy of the subject to this branch of mathematics. It is exactly this viewpoint which distinguishes this text from the previous ones. Although the topics are discrete, our approach is absolutely analytic.

As in continuous mathematical analysis, we first introduce the main concepts such as the definition of a function, power and exponential functions, discrete differentiation and integration, series expansion, complex analytic functions and their integrals, the Cauchy theorem for analytic functions, as well as the harmonic functions in discrete cases. Then we relate these concepts to the theory of discrete differential equations.

The book consists of four chapters and an appendix. Chapter 1 is devoted to preliminary concepts. It can be skipped without any harm to the understanding of what follows in other chapters, except possibly for the definition and the properties of *"arrangement power function"*.

Chapter 2 is concerned with sequences. These sequences are defined by certain recurrence relations which are so called difference equations. To solve these equations, we give several techniques. In particular, we use the discrete transformation for finite sequences and Euler's scheme for infinite ones.

In Chapter 3, we first define the discrete differentiation and integration of a functions as well as discrete ordinary differential equations. Then, according to the nature of the difference equations, we examine them as Cauchy or boundary value problems. At this stage, the reader will see how the mathematical facts will remain valid when passing from continuous cases to discrete ones and vice versa.

The theory of discrete boundary value problems and their connections to the spectral theory of operators will be discussed in Chapter 4. These problems can be interpreted as the discrete analog of the mathematical physics equations in one dimension.

The rest of the book is a long appendix consisting of 10 sections. Except for the three sections; namely an unsolved problem about a cubic root of a complex number, symmetric derivative, and the spectral decomposition of a matrix, all other sections are devoted to the discrete analysis of complex valued functions. Topics included are the discrete partial derivatives of complex valued functions, the Cauchy-Riemann equations, harmonic functions, the Cauchy theorem, Laplace equation, Cauchy-Riemann differential equations, complex integration and above all, the most important concept which brings all these together, is the discrete analog of analytic functions in complex analysis. Finally the author would like to express his gratitude to Dr. Antonio Boccuto for his interesting and promising foreword of the book, to Bentham Science Publishers, in particular to Manager Bushra Siddiqui, for their kindly help, support, and efforts.

F. A. Izadi

December 3, 2009

Toronto Canada

Discrete Calculus By Analogy

1   Preliminary Concepts

1

Abstract. For the purpose of being self-contained, the book starts with some elementary topics. These include the definition of various standard sets, some basic combinatorial definitions and identities, introducing complex numbers and their arithmetic properties, in particular de Moivre formula and the nth roots of a complex number. The case of square root gives rise to an elegant formula for the root of the real number of the form $a \pm \sqrt{b}$. The next two sections are devoted to the definitions and properties of matrices and determinants, the solutions of the linear system of equations as well as the formulas for some special determinants, in particular the Vandermonde determinant. Finally, the closing remark of the chapter is the definition of a function which plays an important role in discrete calculus. Because of its relation to the arrangement formula it is called "the arrangement power function".

1.1   Sets and operations

Throughout the book we use the following notations:

$\mathbb{N}$ = The set of natural numbers

$\mathbb{Z}$ = The set of integers

$\mathbb{Q}$ = The set of rational numbers

$\mathbb{R}$ = The set of real numbers

It is clear that $\mathbb{N} \subset \mathbb{Z} \subset \mathbb{Q} \subset \mathbb{R}$. We use the small letters $a, b, c, \ldots$ to denote the elements of the sets, while the capital letters are used to denote the sets

---

themselves. We write $x \in A$ to indicate that x is an element of the set of A, while $x \notin A$ means that x is not an element of A. The set whose only elements are $a_1, a_2, \ldots, a_n$ is denoted by $\{a_1, a_2, \ldots, a_n\}$, and is called a finite set. We allow the possibility that $a_i = a_j$ even if $i \neq j$. The set whose only element is x is denoted by $\{x\}$, and is called a singleton set. The empty set is denoted by $\emptyset$. We write $A \subset \mathbb{N}$, $B \subset \mathbb{R}$ to indicate that A is a proper subset of $\mathbb{N}$ and B is a proper subset of $\mathbb{R}$. If A and B are subsets of X, we write

$$A \cup B = \{x : x \in A \text{ or } x \in B\}$$
$$A \cap B = \{x : x \in A \text{ and } x \in B\}$$

the terms union and intersection, respectively are used to indicate these combinations of A and B. We write

$$\mathbb{N} \times \mathbb{Q} = \{(x, y) : x \in \mathbb{N} \text{ and } y \in \mathbb{Q}\}$$

to indicate the Cartesian product of $\mathbb{N}$ and $\mathbb{Q}$.

1.2 Combinatorics and binomial identity

Suppose that A is a finite set with elements

$$A = \{a_1, a_2, \ldots, a_n\} .$$

Clearly, combinatorics deals with the number of possible cases on which one can select a k elements from a finite set of n elements regardless of the features of the objects involved. Regarding to this general facts, we give the definition of the two most widely used cases.

(i) An k–permutation of n objects is defined as an ordered arrangement of k of these objects. The number of k–permutation of n objects is denoted by A_n^k and is calculated by the formula

$$A_n^k = n(n-1)\cdots[n-(k-1)] \quad k \leq n \; .$$

In the sequel, we will use this formula to define a specific function which plays an important role in almost every part of the book.

Remark 1. An n–permutation is denoted by P_n. Clearly $P_n = A_n^n = n!$ where, we assume that $0! = 1$.

(ii) An k–combination of n objects is defined as an unordered selection of k of these objects. The number of k–combination is denote by C_n^k and is given by

$$C_n^k = \frac{n!}{k!(n-k)!} \quad k \leq n \; .$$

It is easily seen that

$$C_n^k = \frac{A_n^k}{P_k} \; ,$$

Replacing k by $n-k$ in this expression we see that

$$C_n^{n-k} = \frac{n!}{(n-k)![n-(n-k)]!} = \frac{n!}{(n-k)!k!}$$

or equivalently

$$C_n^{n-k} = C_n^k \; . \tag{1.1}$$

By using the monomials x and y we can define the Newton binomial $(x+y)^n$ as follows

$$(x+y)^n = \sum_{k=0}^{n} C_n^k \, x^k \, y^{n-k} \tag{1.2}$$

where C_n^k are called binomial coefficients.

From the above expression, we see that if $x = y = 1$, then

$$\sum_{k=0}^{n} C_n^k = 2^n \qquad (1.3)$$

i.e., the sum of the binomial coefficients is 2^n. From the property (1.1) it is clear that the binomial coefficients of $(x + y)^n$ in terms of the decreasing powers of x (or y) are symmetric with respect to the middle term (or terms). As we have already assume that $0! = 1$, it is clear from the property (1.1) that the first and last terms have coefficients 1, the second terms from both sides have coefficients n, and so on. Since the Newton binomial is exactly equal to to the following

$$(x + y)(x + y) \cdots (x + y) \qquad n\text{–times} ,$$

and the coefficients of those expressions are all natural numbers, we conclude that the coefficients of the Newton binomial regardless of their apparent forms (i.e., fractions) are all natural numbers. In fact, the product of k consecutive numbers is divisible by $k!$.

1.3   Complex numbers

Suppose that $a \in \mathbb{R}$, $b \in \mathbb{R}$, then $a + ib$, where $i^2 = -1$, is a complex number. Here $a + ib$ stands for $a \cdot 1 + b \cdot i$ where, $a \cdot 1$ (1 is the real unit) is the real part and $b \cdot i$ (i is the imaginary unit) is the imaginary part of the complex number. In this section we deal with the three different forms of the complex numbers. As a matter of fact, the above notation $a + ib$ is the algebraic form of the complex number. As usual, the set of complex numbers is denoted by

$$\mathbb{C} = \{x + iy : x \in \mathbb{R} \quad \text{and} \quad y \in \mathbb{R}, \ i^2 = -1\} .$$

As we use the set $\mathbb{R}$ of real numbers to indicate the points on a real line, we use the set $\mathbb{C}$ of the complex numbers to indicate the points of the plane

$\mathbb{R}^2$. These points are the geometrical form of the complex numbers. In fact, the first coordinates are the real parts and the second coordinates are the imaginary parts of the complex numbers. Finally if we switch the Cartesian coordinates into the polar coordinates, we get the trigonometric form of the complex numbers. To this end, suppose ρ denotes the distance of the complex number $a + ib$ from the origin, and θ is the angle of the vector (a, b) with the positive direction of the x–axis, then using the relations $a = \rho \cos \theta$ and $b = \rho \sin \theta$ we have

$$a + ib = \rho(\cos \theta + i \sin \theta) \tag{1.4}$$

which is the trigonometric form of the complex numbers. Now we are ready to define the algebraic operation of the complex numbers.

It is easily seen that the addition of two complex numbers is not an algebraic sum but a geometric sum. In other words, it is a vector addition. To see this, we give an example. Suppose that we have 3 books and 5 notebooks. If we add 4 books and 3 notebooks to this set, then the resulting set consists of 7 books and 8 notebooks. In order to distinguish the number of books from the notebooks, we use vector addition with the first coordinates as books and the second coordinates as notebooks. From this simple example we see that two complex numbers are equal if and only if the corresponding coordinates are equal to each other. i.e., the real part of the first number is equal to the real part of the second number and similarly for their imaginary parts. This equality of complex numbers in turn implies that a complex number is zero if and only if the real and imaginary parts are both equal to zero. We note that to find a sum of two complex numbers, their algebraic form are more convenient. If $a + ib$ and $c + id$ are two complex numbers, then their algebraic sum is

$$(a + ib) \pm (c + id) = (a \pm c) + i(b \pm d) \,.$$

To find a product of two complex numbers we simply use the multiplication of polynomials by taking into account the fact that $i^2 = -1$. Thus

$$(a + ib)(c + id) = (ac - bd) + i(ad + bc) .$$

Similarly we can show that the quotients of two complex numbers is also a complex number, in fact by assuming $c + id \neq 0$, we have

$$\frac{a + ib}{c + id} = \frac{(a + ib)(c - id)}{(c + id)(c - id)} = \frac{(ac + bd) + i(bc - ad)}{c^2 + d^2} .$$

We see that the result of any binary operation on two complex numbers is a complex number. In this way, the set of complex numbers is closed under arithmetic operations.

Two other important operations are conjugation and taking roots of complex numbers. The conjugate of the complex number is defined by $a - ib$ and denoted by $\overline{a + ib}$, i.e., $\overline{a + ib} = a - ib$, which is again a complex number. Finally we show that any root of a complex number is also a complex number. First we calculate the square root of a complex number. To do this we use the algebraic form of the complex numbers. Suppose that

$$\sqrt{a + ib} = x + iy$$

where a and b are two certain real numbers and x and y are two unknown real numbers which have to be found. Raising to the power of two of both sides we have

$$x^2 - y^2 + 2xyi = a + ib$$

or equivalently

$$\begin{cases} x^2 - y^2 & = a \\ 2xy & = b \end{cases}$$

which is a non-linear system of two unknowns. It can be written in the form

$$\begin{cases} x^2 + (-y^2) & = a \\ x^2(-y^2) & = -b^2/4 . \end{cases}$$

So by Vieta theorem we get the following quadratic equation

$$t^2 - at - b^2/4 = 0 \ .$$

The solutions of this equation from one hand equal to x^2 and $(-y^2)$ and on the other hand equal to

$$\frac{a \pm \sqrt{a^2 + b^2}}{2} \ .$$

Thus we have

$$\begin{cases} x^2 & = \dfrac{\sqrt{a^2 + b^2} + a}{2} \\[2em] y^2 & = \dfrac{\sqrt{a^2 + b^2} - a}{2} \ . \end{cases}$$

Therefore the final solutions are

$$\begin{cases} \sqrt{a + ib} = \pm \left(\sqrt{\dfrac{\sqrt{a^2 + b^2} + a}{2}} + i\sqrt{\dfrac{\sqrt{a^2 + b^2} - a}{2}} \right) & \text{if } b > 0 \\[3em] \sqrt{a + ib} = \pm \left(\sqrt{\dfrac{\sqrt{a^2 + b^2} + a}{2}} - i\sqrt{\dfrac{\sqrt{a^2 + b^2} - a}{2}} \right) & \text{if } b < 0 \\[2em] \ . \end{cases} \qquad (1.5)$$

Corollary. *If a and b are two positive numbers such that $a^2 - b$ is a complete square, i.e., $a^2 - b = c^2$, where c is a real number, then*

$$\sqrt{a \pm \sqrt{b}} = \sqrt{\frac{a + c}{2}} \pm \sqrt{\frac{a - c}{2}} \ . \qquad (1.6)$$

Proof.   In fact by (1.5) we have

$$\sqrt{a \pm \sqrt{b}} = \sqrt{\frac{a + \sqrt{a^2 - b}}{2}} \pm \sqrt{\frac{a - \sqrt{a^2 - b}}{2}}$$

$$= \sqrt{\frac{a + c}{2}} \pm \sqrt{\frac{a - c}{2}} \ .$$

In this way, we get rid of the second square root sign. From this simple example we realize that not only taking the roots of a complex number but also the calculation of its natural power become considerably large by increasing of the order. To get around this difficulty we use the trigonometric form of a complex number. Suppose that

$$Z = r(\cos\theta + i\sin\theta) \ , \quad W = \rho(\cos\varphi + i\sin\varphi)$$

are two complex numbers in trigonometric form, then

$$
\begin{aligned}
ZW &= [r(\cos\theta + i\sin\theta)]\,[\rho(\cos\varphi + i\sin\varphi)] \\
   &= r\rho\,[\cos(\theta + \varphi) + i\sin(\theta + \varphi)]
\end{aligned}
$$

i.e., the module of the product of two complex numbers is the product of their modules and its argument is the sum of their arguments. Similarly we can show that the module of the quotient of two complex numbers is the quotient of their product and its argument is the difference of their arguments, i.e.,

$$\frac{Z}{W} = \frac{r}{\rho}\,[\cos(\theta - \varphi) + i\sin(\theta - \varphi)] \qquad r \neq 0 \ .$$

Using the product formula of two complex numbers inductively, one can easily calculate the product of n complex numbers, in fact if

$$Z_k = r_k\,(\cos\theta_k + i\sin\theta_k) \qquad (k = 1, 2, \ldots, n)$$

are n arbitrary complex numbers, then their product is equal to

$$Z_1 Z_2 \ldots Z_n = r_1 r_2 \cdots r_n\,[\cos(\theta_1 + \cdots + \theta_n) + i\sin(\theta_1 + \cdots + \theta_n)] \ .$$

In particular for $Z_1 = Z_2 = \cdots = Z_n = Z = r(\cos\theta + i\sin\theta)$ we have

$$Z^n = r^2\,[\cos(n\theta) + i\sin(n\theta)] \tag{1.7}$$

which is called de Moivre formula. Using this formula, we can easily find the nth root of a complex number, in fact

$$Z^{1/n} = r^{1/n} \left[\cos\left(\frac{\theta + 2k\pi}{n}\right) + i \sin\left(\frac{\theta + 2k\pi}{n}\right) \right] \quad k \in Z \tag{1.8}$$

where n consecutive values of k, i.e., $k = 0, 1, \ldots, n-1$ suffice, since for other values of k the same roots are repeated. Sometimes (1.8) is also called de Moivre formula.

1.4 Matrix, determinants and their properties

Assuming $m \cdot n$ arbitrary complex numbers in m row in which on every row lie exactly n numbers, we obtain a rectangular array which is called a m by n matrix. In this case we say that the matrix has m rows and n columns. If the number of rows equals to the number of columns, i.e., $m = n$, then the matrix is called a square matrix of order n. The numbers in the matrix are called entries. If all entries on the main diagonal are 1 and elsewhere are zero, then the matrix is called an identity matrix. Usually we use the capital letters $A, B, \ldots$ to denote the matrices and the small letters with double subscripts such as a_{ij} to denote the entries of a matrix, where i, j stand for row and column respectively. In fact we have

$$A = [a_{ij}]_{i,j=1}^{m,n} = \begin{bmatrix} a_{11} & a_{12} & \cdots & a_{1n} \\ a_{21} & a_{22} & \cdots & a_{2n} \\ \vdots & \vdots & & \vdots \\ a_{m1} & a_{m2} & \cdots & a_{mn} \end{bmatrix}$$

Matrices with the same orders can be added or subtracted. To do this we just need to add or subtract the corresponding entries. A matrix with all entries are equal to zero is called the zero matrix. Using this matrix one can define an additive inverse for any matrix, namely the matrix which all entries are additive inverse of the entries of a given matrix. Since the identity matrix is a square matrix, it is possible to define a multiplicative inverse for square matrices. However for non-square matrices we can only define one-sided inverses. As in complex numbers, the binary operations are not the only operations. In other words one can define the transpose or conjugate of a matrix as follows. If $A = [a_{ij}]_{i,j=1}^{m,n}$ is a m by n matrix, then the transpose of A is defined by $[a_{ji}]_{j,i=1}^{n,m}$ and denoted by A^T. Similarly the conjugate of A is defined by $[\overline{a_{ji}}]_{j,i=1}^{n,m}$ and denoted by A^*, where $\overline{a_{ji}}$ stands for conjugate of a_{ji}. More explicitly,

$$
\begin{bmatrix}
a_{11} & a_{12} & \cdots & a_{1n} \\
a_{21} & a_{22} & \cdots & a_{2n} \\
\vdots & \vdots & & \vdots \\
a_{m1} & a_{m2} & \cdots & a_{mn}
\end{bmatrix}^T
=
\begin{bmatrix}
a_{11} & a_{21} & \cdots & a_{m1} \\
a_{12} & a_{22} & \cdots & a_{m2} \\
\vdots & \vdots & & \vdots \\
a_{1n} & a_{2n} & \cdots & a_{mn}
\end{bmatrix}
$$

and

$$
\begin{bmatrix}
a_{11} & a_{12} & \cdots & a_{1n} \\
a_{21} & a_{22} & \cdots & a_{2n} \\
\vdots & \vdots & & \vdots \\
a_{m1} & a_{m2} & \cdots & a_{mn}
\end{bmatrix}^*
=
\begin{bmatrix}
\overline{a_{11}} & \overline{a_{21}} & \cdots & \overline{a_{m1}} \\
\overline{a_{12}} & \overline{a_{22}} & \cdots & \overline{a_{m2}} \\
\vdots & \vdots & & \vdots \\
\overline{a_{1n}} & \overline{a_{2n}} & \cdots & \overline{a_{mn}}
\end{bmatrix}
$$

To define the product of two matrices we need an extra condition, namely the number of columns of the first matrix should equal to the number of rows of the second one. So suppose that

$$
A = [a_{ij}]_{m \times n} \quad , \quad B = [b_{ij}]_{n \times p}
$$

be two such matrices, then define $C = A \cdot B$, where

$$
C = [c_{ij}]_{m \times p} \quad \text{and} \quad c_{ij} = \sum_{k=1}^{n} a_{ik} b_{kj} \; .
$$

Using this definition one can define the multiplicative inverse of the square matrix $A = [a_{ij}]_{n \times n}$ as a matrix $B = [b_{ij}]_{n \times n}$ such that

$$
AB = BA = I
$$

where I is the identity matrix of order n. We usually write:

$$B = A^{-1} \ .$$

Now we are ready to study the determinant of a matrix. In some sense the determinant of a matrix is the numerical character of the matrix. We use the notation $\det A$ or $|A|$ for the determinant of a matrix A. It has the following properties:

a) The interchange of rows with columns does not affect the value of the determinant, i.e., $|A|^T = |A|$.

b) If the rows of the matrix are linearly dependent , then its determinant is equal to zero.

c) If A, B, and C are three $n \times n$ matrices such that all corresponding entries except for the rth row are equal and the entries of rth row of C is equal to the sum of the corresponding entries of A and B, then

$$|C| = |A| + |B| \ .$$

Observe that the first property tells us that there is no difference between rows and columns. In other words any properties of rows is a properties of columns and vice versa. The second property connects the determinant being zero to the relation among the rows of the matrix. In addition, if all entries of a row are equal to zero, or the corresponding entries of two rows are equal or proportional, then the determinant is equal to zero. Finally the third property which has a main difference with the matrix properties, reveals the role of the linear dependence in the determinant.

1.5 Minor, cofactor and calculation of some determinants

By a kth minor of $m \times n$ $(k \le m, k \le n)$ matrix A, we simply mean the determinant of $k \times k$ submatrix of A. Similarly one can define a minor of a square matrix. Suppose that $A = [a_{ij}]_{n \times n}$ be a square matrix, we define the corresponding minor of a_{ij} to be the determinant of $(n-1) \times (n-1)$ submatrix of A which is obtained by eliminating the ith row and j^{th} column of A. Similarly the corresponding cofactor of a_{ij} is the product of the minor of a_{ij} and $(-1)^{i+j}$. This cofactor is denoted by A_{ij}. Hence we have

$$A_{ij} = (-1)^{i+j} \quad \text{minor of } a_{ij} \ .$$

Now the determinant of a matrix $A = [a_{ij}]_{n \times n}$ is defined by

$$|A| = \sum_{i=1}^{n} a_{ij} A_{ij} = \sum_{j=1}^{n} a_{ij} A_{ij} \ .$$

¿From the above property we observe that the product of entries of one row to the cofactors of another is equal to zero. So one can easily show that if $|A| \ne 0$, then

$$A^{-1} = \frac{\text{adj } A}{|A|}$$

where, adj $A = [A_{ij}]_{n \times n}^{T}$.

We conclude this section with the calculation of some important determinants. As it is clear, the determinant of a 2×2 matrix is defined by

$$\begin{vmatrix} a_{11} & a_{12} \\ a_{21} & a_{22} \end{vmatrix} = a_{11}a_{22} - a_{12}a_{21} \ .$$

Suppose that the numbers λ_i $(i = 1, 2, \ldots, n)$ are all distinct. Then we have the following $n \times n$ matrix V

$$V = \begin{bmatrix} 1 & 1 & \cdots & 1 \\ \lambda_1 & \lambda_2 & \cdots & \lambda_n \\ \vdots & \vdots & & \vdots \\ \lambda_1^{n-1} & \lambda_2^{n-1} & \cdots & \lambda_n^{n-1} \end{bmatrix}.$$

The determinant of this matrix is called the Vandermonde determinant and is denoted by

$$D(\lambda_1, \lambda_2, \ldots, \lambda_n) = |V|$$

where

$$D(\lambda_1, \lambda_2, \ldots, \lambda_n) = \prod_{i>j}(\lambda_i - \lambda_j)$$

$$= (\lambda_2 - \lambda_1)(\lambda_3 - \lambda_1)(\lambda_3 - \lambda_2) \cdots (\lambda_n - \lambda_1) \cdots (\lambda_n - \lambda_{n-1}) .$$

As we see from the above formula, if for some i, j, $i \neq j$ but $\lambda_i = \lambda_j$, then the determinant is equal to zero.

Determinant of two triangular matrices. Using the general definition of determinant one can easily show that

$$\det \begin{bmatrix} a_{11} & 0 & 0 & \cdots & 0 \\ a_{21} & a_{22} & 0 & \cdots & 0 \\ \vdots & \vdots & & & \vdots \\ a_{n1} & a_{n2} & \cdots & & a_{nn} \end{bmatrix} = a_{11}a_{22}\cdots a_{nn}$$

and similarly

$$\det \begin{bmatrix} 0 & 0 & \cdots & 0 & 0 & a_{1n} \\ 0 & 0 & \cdots & 0 & a_{2n-1} & a_{2n} \\ 0 & 0 & \cdots & a_{3n-2} & a_{3n-1} & a_{3n} \\ \vdots & \vdots & & \vdots & \vdots & \vdots \\ a_{n1} & a_{n2} & \cdots & a_{nn-2} & a_{nn-1} & a_{nn} \end{bmatrix} = (-1)^{\frac{n(n-1)}{2}} a_{1n}a_{2n-1}a_{3n-1}\cdots a_{n1} .$$

The above matrices are called triangular matrices of type I and type II respectively.

1.6   Solution of a linear system of equations

Consider the linear system of

$$\sum_{j=1}^{n} a_{ij}x_j = b_i \quad (i = 1, \ldots, n) \tag{$*$}$$

where a_{ij} and b_j are given numbers and x_j are some unknowns which have to be found. Suppose that the determinant of the coefficient matrix $A = [a_{ij}]_{n \times n}$ is non-zero. i.e., $|A| = \Delta \neq 0$. Multiplying both sides of ($*$) by A_{ik} i.e., the corresponding cofactor of a_{ik} and summing up with respect to i we get

$$\sum_{i=1}^{n} A_{ik} \sum_{j=1}^{n} a_{ij}x_j = \sum_{i=1}^{n} b_i A_{ik} .$$

Interchanging the sigma notations on the left hand side of the equation we have

$$\sum_{j=1}^{n} \left(\sum_{i=1}^{n} a_{ij} A_{ik} \right) x_j = \sum_{i=1}^{n} b_i A_{ik} .$$

Since

$$\sum_{i=1}^{n} a_{ij} A_{ik} = \delta_{jk}\Delta$$

where δ_{jk} is the Kronecker symbol such that $\delta_{jk} = 0$ if $j \neq k$ and $\delta_{jk} = 1$ if $j = k$, we have

$$\sum_{j=1}^{n} \delta_{jk}\Delta \cdot x_j = \sum_{i=1}^{n} b_i A_{ik}$$

or equivalently

$$\Delta \cdot x_k = \sum_{i=1}^{n} b_i A_{ik} .$$

Denoting the right hand side of the equation by Δ_k we have

$$\Delta_k = \sum_{i=1}^{n} b_i A_{ik}$$

$$= \det \begin{bmatrix} a_{11} & \cdots & a_{1k-1} & b_1 & a_{1k+1} & \cdots & a_{1n} \\ a_{21} & \cdots & a_{2k-1} & b_2 & a_{2k+1} & \cdots & a_{2n} \\ \vdots & & \vdots & \vdots & \vdots & & \vdots \\ a_{n1} & \cdots & a_{nk-1} & b_n & a_{nk+1} & \cdots & a_{nn} \end{bmatrix}$$

whence we have

$$x_k = \Delta_k/\Delta \ ,$$

which is called the Cramer rule for solutions.

Thus we showed that the solution of the system of equations

$$\sum_{j=1}^{n} a_{ij} x_j = b_i \qquad (i = 1, \ldots, n)$$

can be written in the form

$$\Delta \cdot x_k = \Delta_k \qquad (k = 1, \ldots, n) \qquad\qquad (**)$$

where $\Delta = |a_{ij}|_{n \times n}$ is the main determinant of system of equations and

$$\Delta_k = \det \begin{bmatrix} a_{11} & \cdots & a_{1k-1} & b_1 & a_{1k+1} & \cdots & a_{1n} \\ a_{21} & \cdots & a_{2k-1} & b_2 & a_{2k+1} & \cdots & a_{2n} \\ \vdots & & \vdots & \vdots & \vdots & & \vdots \\ a_{n1} & \cdots & a_{nk-1} & b_n & a_{nk+1} & \cdots & a_{nn} \end{bmatrix}$$

are auxiliary determinants.

From the system of equation $(**)$, it is clear that if $\Delta = 0$ and at least one of Δ_k's different from zero, then at least one of the equations $\Delta \cdot x_k = \Delta_k$ fails. Therefore the system of equations $(*)$ has no solution, even if, Δ and all Δ_k's $(k = 1, \ldots, n)$ are simultaneously equal to zero, i.e.,

$$\Delta = \Delta_1 = \cdots = \Delta_n = 0$$

then the system may have still no solution.

As a simple example consider the system

$$\begin{cases} x_1 + 2x_2 + 3x_3 & = 1 \\ 2x_1 + 4x_2 + 6x_3 & = 4 \\ 3x_1 + 6x_2 + 9x_3 & = 2 \end{cases}$$

of 3-equations in 3-unknowns. As we see the columns of the coefficient matrix are pairwise proportional, hence the corresponding columns of Δ_k's are also proportional, so

$$\Delta = \Delta_1 = \Delta_2 = \Delta_3 = 0 \ .$$

However as we observe, this system has clearly no solution.

Now suppose that x_j^0 $(j = 1, \ldots, n)$ is a solution of the system $(*)$. Then we have

$$\sum_{j=1}^{n} a_{ij} x_j^0 = b_i \quad (i = 1, \ldots, n) \ .$$

If we subtract this system from the system $(*)$ side by side, then we get the following system of equations

$$\sum_{j=1}^{n} a_{ij} \left(x_j - x_j^0 \right) = 0 \quad (i = 1, \ldots, n)$$

which is so-called the homogeneous system of equations. If the main determinant of this system is equal to zero, then the system has infinitely many solutions. In fact from the expression

$$x_j - x_j^0 = A_{ij} C \qquad (i, j = 1, \ldots, n)$$

there exists a solution for each value of C, namely

$$x_j = x_j^0 + A_{ij} C \ .$$

Hence we see that all solutions are expressed in terms of a particular solution x_j^0 $(j = 1, \ldots, n)$ and the cofactors of the coefficient matrix of the system.

1.7   Initial value and boundary value problems

Suppose that

$$\left(a_k^p\right)_{k \in \mathbb{N}} \qquad (p = 1, \dots, m)$$

are m arbitrary sequences of real numbers. If there exist m real numbers C_p $(p = 1, \dots, m)$ not all are equal to zero such that

$$\sum_{p=1}^{m} C_p a_k^p = 0 \quad \text{for all } k \in \mathbb{N}$$

then the sequences

$$\left(a_k^1\right), \dots, \left(a_k^m\right)$$

are linearly dependent, otherwise they are called linearly independent. If in a difference equation of order n, the initial terms of the equation is given, then this equation along with the initial conditions is called a Cauchy problem. Similarly if in a differential equation of order n, the values of the function and its derivatives up to $(n-1)^{\text{th}}$ order are given at the initial point, then the equation along with these conditions is called a Cauchy problem.

Finally if in a differential equation of order n, n linear combination of the values of the function and its derivatives up to $(n-1)^{\text{th}}$ order at initial and final points are given, then the equation along with these conditions is called a boundary value problem.

We close this section with the definition of a function which plays an important role in discrete calculus. Because of its relation with the arrangement formula it is called "the arrangement power function".

Suppose that $k \in \mathbb{N}$, $x \in \mathbb{N}$, then the arrangement power function is defined by $A_x^k = \dfrac{x!}{(x-k)!}$ and is denoted by

$$x^{[k]} = A_x^k = \frac{x!}{(x-k)!} \, .$$

This function can be extended as follows:

For $k \in \mathbb{Z}$, $x \in \mathbb{Z}$, define

$$x^{[k]} = \begin{cases} A_x^k & x \geq k \geq 0 \\ 0 & k > x \geq 0 \\ \left(A_{x-k}^{-k}\right)^{-1} & x \geq 0, \ k < 0 \\ \pm\infty & k \leq x < 0 \\ (-1)^k A_{-x+k-1}^k & x < 0, \ k > 0 \ . \end{cases}$$

It can be easily shown that

$$x^{[p+q]} = x^{[p]}(x - p)^{[q]} \quad p, q, x \in \mathbb{Z} \ . \tag{$*$}$$

In fact, if $x \geq p + q > 0$, then the above formula can be proved from the properties of arrangement. Using this formula one can prove analogs of all algebraic identities. Furthermore, by taking $p + q = 0$, we see that $1 = x^{[0]} = x^{[-q]}(x + q)^{[q]}$ which implies the following formula

$$x^{[-q]} = \frac{1}{(x + q)^{[q]}} = \frac{1}{(x + q)(x + q - 1) \cdots (x + 1)} \ .$$

We show that the functions $x^{[k]}$ $(k = 0, 1, \ldots, n)$ are linearly independent. To this end, let us suppose that the numbers C_k $(k = 0, 1, \ldots, n)$ are arbitrary real numbers such that

$$C_0 + C_1 x + C_2 x^{[2]} + \cdots + C_n x^{[n]} = 0 \ .$$

Since this equality holds for any $x \in \mathbb{Z}$, we first assume that $x = 0$. Then substituting $x = 0$ in the right hand side we get $C_0 = 0$ which in turn implies

$$C_1 x + C_2 x^{[2]} + \cdots + C_n x^{[n]} = 0 \ .$$

Dividing this expression by x we have

$$C_1 + C_2(x - 1) + \cdots + C_n(x - 1)^{[n-1]} = 0 \ .$$

Again this equality holds for any $x \in \mathbb{Z}$, in particular taking $x = 1$ gives $C_1 = 0$. Continuing in this procedure, we can show by induction that $C_k = 0$ for $k = 0, 1, \ldots, n$.

As we observe, these functions very look like the power functions in mathematical analysis. This resemblance is not so surprising, in fact, because of their linearly independence, these functions are appropriate basis for all functions in discrete calculus. We shall come back to this point in next chapters.

2   Theory of Sequences

² **Abstract.** In this chapter we discuss sequences which are defined by certain recurrence relations called difference equations. In general, they are divided into two different types; the finite and the infinite sequences. The finite sequences in turn are separated into two different cases; namely the periodic sequences and non-periodic ones. For the solution of a periodic finite sequence, we use finite Fourier series which in this book we refer to as discrete transformation. But to solve the non-periodic finite sequence, it is first necessary to transform it into a periodic sequence which can be done by adding some new equations to the system and then deal with it as a periodic finite sequence. Last section concerned with the solutions of the infinite sequence by using the method of Euler's scheme.

Sequence is a function defined on the set of natural numbers, integers, or arbitrary subsets of these sets on which the range is the subset of real or complex numbers. To determine a sequence it suffices to determine its general term. Suppose that $(a_n)_{n \in \mathbb{Z}}$ be a sequence such that after m consecutive values its terms are repeated, i.e., for all $n \in \mathbb{Z}$, we have $a_{n+m} = a_n$, then this sequence is called an m–periodic sequence. If there is no m with this property, then the sequence is called non-periodic sequence or simply a sequence.

2.1   Discrete transformation

Consider the equation

$$x^n = 1 \qquad (2.1)$$

where, $n \in \mathbb{N}$ is a fixed number. By the de Moivre formula in (1.8) suppose that

$$x_k = \omega^k , \quad k = 0, 1, \ldots, n-1 \tag{2.2}$$

where

$$\omega = \cos \frac{2\pi}{n} + i \sin \frac{2\pi}{n} . \tag{2.3}$$

Using the formulas in (1.8) and (2.3) for $k \in \mathbb{Z}$ in (2.2) we get an n–periodic sequence.

Now suppose that $(a_k)_{k \in \mathbb{Z}}$ is a sequence with period n. Corresponding to the sequence $(a_k)_{k \in \mathbb{Z}}$ we define the sequence $(A_p)_{p \in \mathbb{Z}}$ by the formula

$$A_p = \sum_{k=0}^{n-1} a_k \omega^{-kp} \quad p = 0, 1, \ldots, n-1 . \tag{2.4}$$

Clearly for each $p \in \mathbb{Z}$, $A_{p+n} = A_p$, i.e., the sequence $(A_p)_{p \in \mathbb{Z}}$ is also an n–periodic sequence. This sequence is called the discrete transform of the sequence $(a_k)_{k \in \mathbb{Z}}$. We have the following calculation

$$\sum_{p=0}^{n-1} A_p \omega^{mp} = \sum_{p=0}^{n-1} \omega^{mp} A_p = \sum_{p=0}^{n-1} \omega^{mp} \left(\sum_{k=0}^{n-1} a_k \omega^{-kp} \right)$$

$$= \sum_{k=0}^{n-1} a_k \sum_{p=0}^{n-1} \omega^{(m-k)p} = \sum_{k=0}^{n-1} a_k \left(1 + \omega^{m-k} + \cdots + \omega^{(n-1)(m-k)} \right)$$

$$= \sum_{\substack{k=0 \\ k \neq m}}^{n-1} a_k \left(1 + \omega^{m-k} + \cdots + \omega^{(n-1)(m-k)} \right) + n \cdot a_m$$

$$= \sum_{\substack{k=0 \\ k \neq m}}^{n-1} a_k \left\{ \frac{1 - \omega^{n(m-k)}}{1 - \omega^{(m-k)}} \right\} + n a_m .$$

Since $\omega^n = 1$, the expression inside the bracket is equal to zero, hence we get the inverse transformation as

$$a_m = \frac{1}{n} \sum_{p=0}^{n-1} A_p \omega^{mp} \quad m = 0, 1, \ldots, n-1 . \tag{2.5}$$

This two transformations are called corresponding discrete transformations.

2.2   Periodic finite sequence scheme

Let us suppose that the n–periodic sequence $(a_k)_{k\in\mathbb{Z}}$ satisfies in the difference equation

$$\sum_{j=0}^{m}\alpha_j a_{k+j}=b_k \quad k=0,1,\ldots,n-1 \tag{2.6}$$

where m is a fixed positive integer such that $m<n$ and α_j $(j=0,1,\ldots,m)$, b_k $(k=0,1,\ldots,n-1)$ are certain real numbers. We wish to determine the sequence $(a_k)_{k\in\mathbb{Z}}$. To this end, we use the transformation (2.4) in equations (2.6). We first multiply both sides of (2.6) by ω^{-kp} and then add them up with respect to the index k from 0 to $(n-1)$, hence

$$\sum_{k=0}^{n-1}\omega^{-kp}\sum_{j=0}^{m}\alpha_j a_{k+j}=\sum_{k=0}^{n-1}\omega^{-kp}b_k \ . \tag{2.7}$$

Clearly the right hand side of (2.7) is the discrete transform of the sequence $(b_k)_{k\in\mathbb{Z}}$. We denote it by B_p. By interchanging the sigma notations on the left-hand side and doing some easy manipulations we have

$$\sum_{k=0}^{n-1}a_{k+j}\omega^{-kp}=\omega^{jp}\sum_{k=0}^{n-1}a_{k+j}\omega^{-(k+j)p}$$

$$=\omega^{jp}\sum_{\ell=j}^{n-1+j}a_\ell\omega^{-\ell p} \ .$$

Since the sequence $(a_k)_{k\in\mathbb{Z}}$ is n–periodic and $\omega^{-np}=1$, we see that

$$a_{n+s}\omega^{-(n+s)p}=a_{n+s}\omega^{-np}\omega^{-sp}=a_s\omega^{-sp} \quad s=0,1,\ldots,n-1+j \ .$$

Therefore we can rewrite the above summation as

$$\sum_{k=0}^{n-1} a_{k+j}\omega^{-kp} = \omega^{jp}\left\{\sum_{\ell=j}^{n-1} a_\ell\omega^{-\ell p} + \sum_{\ell=n}^{n-1+j} a_\ell\omega^{-\ell p}\right\}$$

$$= \omega^{jp}\left\{\sum_{\ell=j}^{n-1} a_\ell\omega^{-\ell p} + \sum_{\ell=0}^{j-1} a_\ell\omega^{-\ell p}\right\}$$

$$= \omega^{jp}\left\{\sum_{\ell=j}^{n-1} a_\ell\omega^{-\ell p} + \sum_{\ell=0}^{j-1} a_\ell\omega^{-\ell p}\right\}$$

$$= \omega^{jp}\sum_{\ell=0}^{n-1} a_\ell\omega^{-\ell p} \ .$$

The above formulas and (2.7) altogether lead to the following simple equation for $(A_p)_{p\in\mathbb{Z}}$

$$\sum_{j=0}^{m} \alpha_j\omega^{jp} A_p = B_p \qquad p = 0, 1, \ldots, n-1 \ . \tag{2.8}$$

¿From (2.8) it is clear that if the condition

$$F_p = \sum_{j=0}^{m} \alpha_j\omega^{jp} \neq 0 \tag{2.9}$$

holds, then for the sequence $(A_p)_{p\in\mathbb{Z}}$ we get the following solution

$$A_p = B_p/F_p \qquad p = 0, 1, \ldots, n-1 \ . \tag{2.10}$$

Finally, to solve the equation (2.6) we first substitute the value of A_p from (2.10) into (2.5) and then the value of B_p into (2.10) and obtain the following formula for a_m

$$a_m = \frac{1}{n} \sum_{p=0}^{n-1} \omega^{mp}(B_p/F_p)$$

$$= \frac{1}{n} \sum_{p=0}^{n-1} \left\{ \sum_{k=0}^{n-1} \omega^{-kp} b_k / F_p \right\}$$

$$= \frac{1}{n} \sum_{k=0}^{n-1} b_k \left\{ \sum_{p=0}^{n-1} \omega^{(m-k)p} / F_p \right\}.$$

We can rewrite the above formula in the closed form

$$a_m = \sum_{k=0}^{n-1} g(m-k)b_k \qquad m = 0, 1, \ldots, n-1 \qquad (2.11)$$

where

$$g(m-k) = \frac{1}{n} \sum_{p=0}^{n-1} \omega^{(m-k)p} / F_p . \qquad (2.12)$$

Note that the expression in (2.11) i.e., the solution of the equation in (2.6) is valid only if (2.9) holds.

Remark 1. If for some value of p, say, $p = p_0$, we have $F_{p_0} = 0$, then for the existence of solution of (2.6) from (2.8) it is necessary that $B_{p_0} = 0$. Otherwise, if $B_{p_0} \neq 0$, then the equation (2.8) does not hold at least for $p = p_0$. This means that there is no A_{p_0} and this in turn implies that there exists no solution for the equation (2.6).

Remark 2. If for some value of p, say, $p = p_0$ we have $F_{p_0} = B_{p_0} = 0$, then for each fixed number A_{p_0} the equation (2.8) does hold. This means that the equation (2.6) has a solution (2.11) for each fixed number A_{p_0}. In other words the equation (2.6) has infinitely many solutions.

Example.   Let $m = 1$, $\alpha_0 = -1$, $\alpha_1 = 1$, then the condition (2.9) does not hold for $p = 0$. Because

$$F_0 = \alpha_0 + \alpha_1 = 0 \ .$$

From Remark 1 we observe that to have a solution (if it exists, it will be n–periodic) we have to have

$$B_0 = \sum_{k=0}^{n-1} b_k = 0 \ .$$

In face, if we rewrite the above system in open form by using the conditions $a_n = a_0$ and $\sum_{k=0}^{n-1} b_k = 0$, then for the sequence $(a_\ell)_{\ell \in \mathbb{Z}}$ we get the solution

$$a_\ell = a_0 + \sum_{k=0}^{\ell-1} b_k \quad \ell = 1, \ldots, n-1, n \ ,$$

where a_0 is a fixed number.

2.3   Non-periodic finite sequence scheme

Again we consider the system of equations

$$\sum_{j=0}^{m} \alpha_j a_{k+j} = b_k \qquad k = 0, 1, \ldots, n-1 \ ,$$

where $(a_k)_{k \in \mathbb{Z}}$ is a non-periodic finite sequence which has to be determined. The main difference between this case and the equations in (2.6) is that in addition to α_j $(j = 0, 1, \ldots, m)$ and b_k $(k = 0, 1, \ldots, n-1)$, the first m consecutive terms i.e., a_k $(k = 0, 1, \ldots, m-1)$ of the sequence $(a_k)_{k \in \mathbb{Z}}$ are also known. Therefore, to determine this sequence completely it suffices to determine its remaining terms, i.e., a_k $(k = m, m+1, \ldots, n-1+m)$. Since the number of a_k's in the system is equal to $n + m - 1$, to have a square system of equations, we add m new equations to this system. In fact instead of system of equations in (2.6) we consider

$$\sum_{j=0}^{m} \alpha_j a_{k+j} = b_k \quad k = 0, 1, \ldots, n+m-1 \tag{2.13}$$

where the new terms of the sequence $(a_k)_{k \in \mathbb{Z}}$ satisfy the following relations

$$a_{n+m} = a_0, \quad a_{n+m+1} = a_1, \ldots, a_{n+m+m-1} = a_{m-1} .$$

In this way, we get an $(n+m)$–periodic sequence. Now in order to solve this system we first need to determine the new b_k $(k = n, n+1, \ldots, n+m-1)$ in (2.13). This can actually be done by using the given a_k $(k = 0, 1, \ldots, m-1)$. To do this, we calculate the formula (2.11) for the equations (2.13), so

$$a_q = \sum_{k=0}^{n+m-1} g(q-k)b_k \quad q = 0, 1, \ldots, n+m-1 \tag{2.14}$$

where

$$g(q-k) = \frac{1}{n+m} \sum_{p=0}^{n+m-1} \omega^{(q-k)p} / F_p$$

and

$$F_p = \sum_{j=0}^{m} \alpha_j \omega^{jp} .$$

Next by using (2.14) for b_k $(k = n, n+1, \ldots, n+m-1)$ we form the system of equations

$$\sum_{k=n}^{n+m-1} g(q-k)b_k = a_q - \sum_{k=0}^{n-1} g(q-k)b_k \quad q = 0, 1, \ldots, m-1 . \tag{2.15}$$

Now in order to have a solution, by Cramer's rule it is sufficient that the coefficient matrix i.e.,

$$G = \begin{pmatrix} g(-n) & g(-n-1) & \cdots & g(-n-m+1) \\ g(1-n) & g(-n) & \cdots & g(-n-m+2) \\ \vdots & \vdots & & \vdots \\ g(m-1-n) & g(m-2-n) & \cdots & g(-n) \end{pmatrix} \tag{2.16}$$

has non-zero determinant. In fact under the condition $|G| \neq 0$ the system of (2.15) has a unique solution. By substituting the solution of this system into (2.14) we get the solution of the $(n+m)$–periodic sequence $(a_k)_{k \in \mathbb{Z}}$ of (2.6).

2.4   Infinite sequence scheme

Suppose that the sequence $(a_k)_{k \in \mathbb{N}}$ satisfies in the following homogeneous system of equations

$$\sum_{j=0}^{m} \alpha_j a_{k+j} = 0 \qquad k \in \mathbb{N} \tag{2.17}$$

where, α_j $(j = 0, 1, \ldots, m)$ and a_k $(k = 1, 2, \ldots, m)$ are certain numbers and a_k $(k > m)$ are unknown numbers yet to be determined. We wish to seek for the solution of the system (2.17) in the form

$$a_k = \lambda^k \tag{2.18}$$

where, $\lambda \in \mathbb{C}$ is unknown. Without loss of generality we suppose that $\lambda \neq 0$. To find λ we first substitute (2.18) into (2.17) and then divide both sides of the equation by λ^k, so we get the following characteristic equation:

$$F(\lambda) = \sum_{j=0}^{m} \alpha_j \lambda^j = 0 . \tag{2.19}$$

If we denote the roots of this equation by λ_p $(p = 1, \ldots, m)$, and assume that all of these roots are pairwise distinct and also different from zero, i.e.,

$$\lambda_p \neq 0 , \quad \text{and} \quad \lambda_p \neq \lambda_q \text{ if } p \neq q , \quad (p, q = 1, \ldots, m) \tag{2.20}$$

then we get the following particular solutions

$$\left(\lambda_p^k\right)_{k \in \mathbb{N}} \quad p = 1, \ldots, m .$$

Since these solutions are linearly independent [condition (2.20)], for arbitrary values of C_p $(p = 1, \ldots, m)$ the expression

$$a_k = \sum_{p=1}^{m} C_p \lambda_p^k \qquad (2.21)$$

is a general solution of (2.17), where p is the summation index and k is the power of λ_p.

Now in order to find the coefficients C_p we use the values of a_k $(k = 1, \ldots, m)$. Therefore form (2.21) we obtain the following linear system of equations

$$\lambda_1 C_1 + \cdots + \lambda_m C_m = a_1$$
$$\lambda_1^2 C_1 + \cdots + \lambda_m^2 C_m = a_2$$
$$\vdots$$
$$\lambda_1^m C_1 + \cdots + \lambda_m^m C_m = a_m .$$

Since the determinant of the coefficients matrix of this system is the Vandermonde determinant and this determinant is different from zero because of (2.20), so by Cramer's rule the system has a unique solution. By substituting this solution in (2.21), we obtain the required solution of (2.17).

Remark 3.   Now suppose that the roots of the characteristic equation in (2.19) are repeated. For convenience we assume that this equation has only one root say, $\lambda = \lambda_0$. Since the repeating degree of this root is m, then the particular solutions of (2.17) will be a linearly dependent set of sequence $\{\lambda_0^k\}_{k \in \mathbb{N}}$. In order to have a linearly independent set, we multiply λ_0^k by some different powers of k namely,

$$k^s \lambda_0^k \qquad s = 0, 1, \ldots, m - 1 .$$

Thus the general solution of (2.17) is

$$a_k = \sum_{p=0}^{m-1} C_p \lambda_0^k k^p = \lambda_0^k \sum_{p=0}^{m-1} C_p k^p .$$

where the coefficients C_p can be determined as the same way in previous paragraph. If $\lambda_0 = 0$, then the system of (2.17) can be easily solved, but if $\lambda_0 = 1$, then the independent particular solutions are

$$k^s \qquad s = 0, 1, \ldots, m - 1.$$

2.5   Solution of a non-homogeneous difference equation

Consider the equation

$$\sum_{j=0}^{m} \alpha_j a_{k+j} = b_k \qquad k \geq 0 \tag{2.22}$$

where α_j ($j = 0, 1, \ldots, m$), a_k ($k = 0, 1, \ldots, m - 1$), and b_k ($k \geq 0$) are assumed to be known, and a_k ($k \geq m$) denote the unknowns which are to be determined. Our objective in this section is to find a particular solution of (2.22). For this purpose, we will examine a few special cases.

Case I.   Suppose that the right-hand side of (2.22) be a polynomial of degree r in k, i.e.,

$$b_k = \sum_{p=0}^{r} \beta_p k^p .\tag{2.23}$$

In this case it is convenient to seek a particular solution of (2.22) in the following form

$$a_k = \sum_{p=0}^{r} C_p k^p .\tag{2.24}$$

So if we substitute the expressions of (2.23) and (2.24) into (2.22), we get the following identity, in terms of k,

$$\sum_{j=0}^{m} \alpha_j \sum_{p=0}^{r} C_p(k+j)^p = \sum_{p=0}^{r} \beta_p k^p \ .$$

After some simple manipulations such as expanding the binomial expression on the left-hand side, rearranging the terms of resulting expression in terms of k and finally equating the corresponding coefficients on both sides we get a system of equation in terms of C_p, i.e.,

$$\sum_{p=q}^{r} C_p \sum_{j=0}^{m} \alpha_j C_p^q j^{p-q} = \beta_q \qquad q = 0, 1, \ldots, r \ . \tag{2.25}$$

Now we rewrite this system in the following open form

$$\sum_{j=0}^{m} \alpha_j \left\{ C_0 + C_1^0 j C_1 + C_2^0 j^2 C_2 + \cdots + C_r^0 j^r C_r \right\} = \beta_0$$

$$\sum_{j=0}^{m} \alpha_j \left\{ C_1^1 + C_2 j C_2 + \cdots + C_r^1 j^{r-1} C_r \right\} = \beta_1$$

$$\vdots$$

$$\sum_{j=0}^{m} \alpha_j \left\{ C_r \right\} = \beta_r \ .$$

As we observe, the coefficient matrix of this system is an upper triangular matrix where all the entries on the main diagonal are equal to $\sum_{j=0}^{m} \alpha_j$. Thus under the condition

$$\sum_{j=0}^{m} \alpha_j \neq 0 \tag{2.26}$$

the determinant of the coefficient matrix, i.e., $\left(\sum_{j=0}^{m} \alpha_j \right)^{r+1}$ will be different from zero and henceforth, according to Cramer's rule the system has a unique solution. By substituting this solution into (2.24) we find a particular solution for (2.22).

Remark 4. Since its coefficients matrix is upper triangular, the system of (2.25) can be easily solved. Because if $q = r$, then $p = q = r$ and the last equation of the system is

$$C_r \sum_{j=0}^{m} \alpha_j = \beta_r \ .$$

Keeping in mind the condition (2.26), we find C_r. By substituting this value in the preceding equation we get the following for C_{r-1}

$$C_{r-1} \sum_{j=0}^{m} \alpha_j + C_r \sum_{j=0}^{m} \alpha_j r \cdot j = \beta_{r-1} \ .$$

By continuing in this procedure, all C_r's are uniquely found. In fact, we solved the system of (2.25) by the Gauss method.

Remark 5. If the condition (2.26) does not hold, i.e.,

$$\sum_{j=0}^{m} \alpha_j = 0$$

and the right-hand side of (2.22) is in the form of (2.23), then we seek the solution in the form

$$a_k = \sum_{p=0}^{r} C_p k^{p+1} \ .$$

By repeating the same argument, instead of the system of (2.25) for C_p we find the following system

$$C_r \sum_{j=0}^{m} \alpha_j = 0$$

$$\sum_{j=0}^{m} \alpha_j \sum_{p=r-1}^{r} C_p C_p^{r-1} j^{p-r+1} + \sum_{j=0}^{m} \alpha_j j C_r = \beta_r$$

$$\sum_{j=0}^{m} \alpha_j \sum_{p=s}^{r} C_p C_p^{s} j^{p-s} \qquad\qquad 0 \leq s < r-1 \quad (2.27)$$

$$+ \sum_{j=0}^{m} \alpha_j j \sum_{p=s+1}^{r} C_p C_p^{s+1} j^{p+s+1} = \beta_{s+1}$$

$$\sum_{j=0}^{m} \alpha_j j \sum_{p=0}^{r} C_p j^{p} = \beta_0$$

We observe from this system that the first equation is equivalent to the negation of (2.26). The second equation reduces to

$$C_r(r+1) \sum_{j=0}^{m} \alpha_j j = \beta_r \; .$$

From this it is clear that if instead of (2.26) we have

$$\sum_{j=0}^{m} j \alpha_j \neq 0 \qquad\qquad (2.28)$$

then like the system of (2.25) the system (2.27) has also a unique solution. This solution can also be found by the Gauss method. If none of the conditions (2.26) and (2.28) hold, then one can handle the solution by increasing the degree of k in (2.24), however in this case the expressions involved in our discussions become more complicated.

Case II.   Suppose that the right-hand side of (2.22) is in the form

$$b_k = b \lambda_0^k \qquad\qquad (2.29)$$

where, $\lambda_0 \neq 0$, $\lambda_0 \neq 1$. (In fact, we have already examined these cases. The equation with $\lambda_0 = 0$ is a homogeneous equation and with $\lambda_0 = 1$ is a non-homogeneous equation with the right-hand side as a polynomial.) In this case we seek the solution in the form

$$a_k = C\lambda_0^k \tag{2.30}$$

where C is an arbitrary fixed constant. Now by substituting (2.29) and (2.30) in (2.22) and dividing both sides of the equation by λ_0^k we get the equation

$$\sum_{j=0}^{m} \alpha_j C \lambda_0^j = b \tag{2.31}$$

in terms of C. From (2.31) it is clear that if $\lambda = \lambda_0$ is not a root of the corresponding characteristic equation of (2.22), then

$$F(\lambda_0) = \Sigma \alpha_j \lambda_0^j \neq 0 \tag{2.32}$$

gives us the constant C by which the particular solution of (2.22) become

$$a_k = (b/F(\lambda_0))\, \lambda_0^k \ .$$

If (2.32) does not hold, i.e., λ_0 is a root of the characteristic equation and the right-hand side of (2.22) is in the form of (2.29) then the particular solution must be in the form

$$a_k = Ck\lambda_0^k \ .$$

By continuing the same procedure and keeping in mind that the condition (2.32) does not hold, the equation for C is

$$C\sum_{j=0}^{m} \alpha_j j \lambda_0^j = b \ .$$

In fact, if $\sum_{j=0}^{m} \alpha_j j \lambda_0^j \neq 0$, then C can be uniquely determined. In this way, we established the particular solution of (2.22).

We return one last time to non-homogeneous equation of (2.22). If we denote a particular solution of this equation by $\tilde{a}_k$, then by change of the variable

$$a_k = \tilde{a}_k + a_k^0$$

for a_k^0 we have the homogeneous equation

$$\sum_{j=0}^{m} \alpha_j a_{k+j}^0 = 0 \ .$$

In order to find the general solution of the non-homogeneous equation we examine the homogeneous equation. In this way the general solution of the non-homogeneous equation is obtained by using the general solution of the homogeneous equation plus a particular solution of the non-homogeneous equation. According to linearity property of (2.22) if the right-hand side i.e., b_k is in the form

$$b_k = \sum_{p=0}^{s} \lambda_p^k \sum_{q=0}^{h_p} \beta_q k^q \tag{2.33}$$

then by taking the right-hand side as $\lambda_p^k \sum_{q=0}^{h_p} \beta_q k^q$ we can express the particular solution as a linear combination of these rather simple particular solutions.

Remark 6. For establishing the general solution of (2.22) we could also use the Lagrange's method of the variation of parameters, but because of some complicated expression involved in this method it will be postponed until next chapters.

In this stage, we examine the general solution of a finite sequence on which the end points are assumed to be known.

Let $\{a_k\}_{k=0}^{n}$ satisfies in the

$$\sum_{j=0}^{2} \alpha_j a_{k+j} = b_k \qquad k = 0, 1, \ldots, n-1 \qquad (2.34)$$

with the boundary value conditions

$$a_0 = a_n = 0 . \qquad (2.35)$$

If (2.34) holds for $k = n-1$ and $k = n$, we can consider the sequence $\{a_k\}_{k=0}^{n}$ as a periodic sequence and use the discrete transformation in Section 2.1 to get the solution. Regardless of the condition (2.35) if we suppose that

$$a_{n+1} = a_1 , \qquad a_{n+2} = a_2$$

then validity of (2.34) for $k = 0, 1, \ldots, n$ is trivial. (b_{n-1} and b_n are certain constants which can be determined from the boundary condition (2.35).) Using the same arguments as in Section 2.4 we can find the expression

$$a_q = \sum_{k=0}^{n} g(q-k)b_k \qquad (2.36)$$

where $g(q-k) = \frac{1}{n} \sum_{p=0}^{n} \frac{\varepsilon^{(q-k)p}}{F(p)}$ and $F(p) = \sum_{j=0}^{2} \alpha_j \varepsilon^{jp} \neq 0$. Now for b_{n-1} and b_n we have the following system of equations

$$\begin{cases} g(1-n)b_{n-1} + g(-n)b_n & = -\sum_{k=0}^{n-2} g(-k)b_k \\ g(1)b_{n-1} + g(0)b_n & = -\sum_{k=0}^{n-2} g(n-k)b_k . \end{cases}$$

By substituting the solution of this system in (2.36) we obtain the solutions of (2.34) and (2.35).

2.6   Matrix representation of difference equations

Consider the system of equations

$$\sum_{j=0}^{m} \alpha_j a_{k+j} = b_k \qquad k \geq 0 \tag{2.37}$$

where α_j $(j = 0, 1, \ldots, m)$ and b_k $(k \geq 0)$ are assumed to be known. In order to write this system in a matrix form we arrange the following notations

$$a_k = a_k^0$$

$$a_{k+1} = a_k^1 = a_{k+1}^0$$

$$a_{k+2} = a_k^2 = a_{k+1}^1 \tag{2.38}$$

$$\vdots$$

$$a_{k+m-1} = a_k^{m-1} = a_{k+1}^{m-2}$$

Substituting these expressions in (2.37), we get

$$\alpha_0 a_k^0 + \alpha_1 a_k^1 + \alpha_2 a_k^2 + \cdots + \alpha_{m-1} a_k^{m-1} + \alpha_m a_{k+1}^{m-1} = b_k \ . \tag{2.39}$$

Now considering (2.38) and (2.39) simultaneously, we get

$$\begin{bmatrix} a_{k+1}^0 \\ a_{k+1}^1 \\ \vdots \\ a_{k+1}^{m-1} \end{bmatrix} = \begin{bmatrix} 0 & 1 & 0 & \cdots & 0 & 0 \\ 0 & 0 & 1 & \cdots & 0 & 0 \\ \vdots & \vdots & \vdots & & \vdots & \vdots \\ 0 & 0 & 0 & \cdots & 0 & 1 \\ \beta_0 & \beta_1 & \beta_2 & \cdots & \beta_{m-2} & \beta_{m-1} \end{bmatrix} \begin{bmatrix} a_k^0 \\ a_k^1 \\ \vdots \\ a_k^{m-2} \\ a_k^{m-1} \end{bmatrix} + \begin{bmatrix} 0 \\ 0 \\ \vdots \\ 0 \\ b \end{bmatrix}$$

where $b = b_k/\alpha_m$ and $\beta_i = -\alpha_i/\alpha_m$ $(i = 0, 1, \ldots, m-1)$. We can write this matrix equation in the closed form

$$A_{k+1} = \alpha A_k + F_k \tag{2.40}$$

where

$$
\alpha = \begin{bmatrix}
0 & 1 & 0 & \cdots & 0 & 0 \\
0 & 0 & 1 & \cdots & 0 & 0 \\
\vdots & \vdots & \vdots & & \vdots & \vdots \\
0 & 0 & 0 & \cdots & 0 & 1 \\
\beta_0 & \beta_1 & \beta_2 & \cdots & \beta_{m-2} & \beta_{m-1}
\end{bmatrix}
$$

is a square matrix of $m \times m$ and

$$
A_k = \begin{bmatrix} a_k^0 \\ a_k^1 \\ \vdots \\ a_k^{m-1} \end{bmatrix} \qquad
F_k = \begin{bmatrix} 0 \\ 0 \\ \vdots \\ b_k/\alpha_m \end{bmatrix} \tag{2.41}
$$

are column vectors of $m \times 1$. We should stress once more that the expression (2.40) is the matrix representation of (2.37). By solving this system, we get

$$
A_k = \sum_{j=1}^{k-1} \alpha^j F_{k-1-j} + \alpha^k A_0 \quad k > 0 \tag{2.42}
$$

where A_0 is an arbitrary column matrix.

3   Discrete Calculus

[3] **Abstract.** Firstly, we define the discrete differentiation and integration of functions and their arithmetic properties. By using these properties, we can easily calculate many important summation formulas in particular a closed formula for the sum of kth powers of the first n consecutive positive integers, an expression for the discrete gamma function, as well as a formula for the number of regions in which x hyperspaces in general position, divide the n-dimensional space R^n. It turns out that this number can be given as a function of x expressible as a Taylor expansion at the origin. Secondly, by introducing an analog of the exponential function, we treat the difference equations as a discrete differential equations. For example, we solve the arithmetic progression as a first order and Fibonacci sequence as a second order Cauchy discrete differential equation problems. Finally, in the remaining three sections, we deal with the methods of the variation of parameters, the solution of the Cauchy problem for non-homogeneous differential equations, as well as the boundary value problem both for homogeneous and non-homogeneous differential equations of second order in discrete cases.

In this chapter the domain of the functions involved are assumed to be the discrete set $\mathbb{Z}$ or any arbitrary subsets of this set.

3.1   Discrete derivative and its properties

Let us suppose that $y = f(x)$ be a function defined on integers. Then the range of f is

$$\{\ldots, f(-2), f(-1), f(0), f(1), f(2), \ldots\} \ .$$

For convenience, we use the notations $f(n)$ and f_n interchangeably throughout the book.

Definition. Suppose that $f : A \to \mathbb{R}$ is given where $A \subseteq \mathbb{Z}$. Then the discrete derivative of f at $x \in A$ is defined to be

$$f(x+1) - f(x) . \qquad (3.1)$$

Using the standard notation $f'(x)$ for this expression we have

$$f'(x) = f(x+1) - f(x) .$$

Remark. In difference calculus, the expression $f(x+1) - f(x)$ is referred to as the difference of $f(x)$, and symbolically is denoted by $\Delta f(x)$. Here we use the terminology "discrete derivative" because it is nothing else but an immediate consequence of usual continuous differentiation formula, in fact if

$$Df(x) = \lim_{h \to 0} \frac{f(x+h) - f(x)}{h}$$

then by setting $h = 1$ we have

$$\Delta f(x) = \lim_{h=1} \frac{f(x+1) - f(x)}{1}$$
$$= f(x+1) - f(x) .$$

In the sequel, we will see that each continuous concept has, in a certain sense, a discrete counterpart which plays almost the same role as continuous one.

An immediate consequence of this definition is that any difference equation can be regarded as a differential equation with discrete domain. Let us examine a few simple examples.

Example 1. *Arithmetic progression.* By definition the sequence $(a_n)_{n\in\mathbb{N}}$ forms an arithmetic progression, if for each $n \geq 1$ we have

$$a_{n+1} = a_n + d$$

where d and a_1 are certain arbitrary constants. By using the notation $a(n)$ and replacing n by x we have

$$a(x+1) - a(x) = d \ .$$

Thus according to the definition (3.1) we get the differential equation

$$a'(x) = d \qquad x \geq 1 \ . \tag{3.2}$$

It is clear that

$$a(1) = a_1 \ . \tag{3.3}$$

We observe that this problem by what we said in Section 1.7 gives rise to a Cauchy problem where the expression (3.3) is the initial condition for the differential equation.

Example 2. *Fibonacci sequence.* As is well-known the Fibonacci sequence is defined by the following difference equation

$$a_{n+2} = a_{n+1} + a_n \qquad n \geq 1$$

where a_1 and a_2 are certain constant numbers. As in arithmetic progression replacing a_n by $a(x)$ gives us

$$a(x+2) = a(x+1) + a(x)$$

or equivalently

$$a'(x+1) = a(x) \ .$$

Adding $-a'(x)$ on both sides we get

$$a'(x+1) - a'(x) = a(x) - a'(x)$$

or equivalently

$$(a(x+1) - a(x))' = a(x) - a'(x)$$

whence by regarding $(a'(x))' = a''(x)$ we obtain the differential equation

$$a''(x) + a'(x) - a(x) = 0 \qquad x \geq 1 \tag{3.4}$$

with the initial condition

$$\begin{cases} a(1) = a_1 \\ a'(1) = a_2 - a_1 \ . \end{cases} \tag{3.5}$$

In other words, the solution of the Fibonacci sequence gives rise to the Cauchy equations of (3.4) and (3.5).

We are now going to state main properties of the discrete derivative.

Property I. *Additive property.* Suppose that $y(x) = u(x) + v(x)$, then

$$y'(x) = u'(x) = v'(x) \ .$$

In fact, we have

$$y'(x) = y(x+1) - y(x) = [u(x+1) + v(x+1)] - [u(x) + v(x)]$$
$$= [u(x+1) - u(x)] + [v(x+1) - v(x)]$$
$$= u'(x) + v'(x) \ .$$

That is,

$$(u(x) + v(x))' = u'(x) + v'(x) \ . \tag{3.6}$$

This can be easily extended to any finite number of terms and even to infinite number if the order of the terms involved may be altered without affecting its sum.

Property II. *The commutative property.* Let c be a fixed real number, and $y(x)$ is a discrete function, then

$$(cy(x))' = cy(x+1) - cy(x) = c\left(y(x+1) - y(x)\right)$$
$$= cy'(x) .$$

That is the derivative of the function cy is the c times the derivative of $y(x)$.

Property III. *Multiplicative property.* Suppose that $y(x) = u(x)v(x)$, then

$$y'(x) = u'(x)v'(x) + u'(x)v(x) + v'(x)u(x) .$$

In the same as above we have

$$\begin{aligned}
y'(x) &= y(x+1) - y(x) = [u(x+1)v(x+1)] - [u(x)v(x)] \\
&= u(x+1)v(x+1) - u(x)v(x+1) + u(x)v(x+1) - u(x)v(x) \\
&= v(x+1)\left[u(x+1) - u(x)\right] + u(x)\left[v(x+1) - v(x)\right] \\
&= v(x+1)u'(x) + u(x)v'(x) \\
&= \left[v(x+1) - v(x) + v(x)\right]u'(x) + u(x)v'(x) \\
&= \left[v'(x) + v(x)\right]u'(x) + u(x)v'(x) \\
&= v'(x)u'(x) + v(x)u'(x) + u(x)v'(x) .
\end{aligned}$$

That is,

$$(u(x)v(x))' = u'(x)v'(x) + u'(x)v(x) + u(x)v'(x) .$$

Property IV. *Division property.*   Let $y(x) = \frac{u(x)}{v(x)}$, then

$$y'(x) = \frac{u(x+1)}{v(x+1)} - \frac{u(x)}{v(x)} = \frac{v(x)u(x+1) - u(x)v(x+1)}{v(x+1)v(x)}$$

$$= \frac{v(x)u(x+1) - u(x)v(x) + u(x)v(x) - u(x)v(x+1)}{v(x)v(x+1)}$$

$$= \frac{v(x)\left[u(x+1) - u(x)\right] - u(x)\left[v(x+1) - v(x)\right]}{v(x)v(x+1)}$$

$$= \frac{v(x)u'(x) - v'(x)u(x)}{v(x)v(x+1)}$$

which is almost the same as the corresponding formula in the differential calculus.

Property V. *Compositive property.*   Consider the composite function $y = u(v(x))$, where $x \in A \subseteq \mathbb{Z}$ and $v(x) \in B \subseteq \mathbb{Z}$. We point out that if $B \not\subseteq \mathbb{Z}$, then the function is not a discrete function. Suppose that V is a strictly increasing function, namely,

$$V(x+1) > V(x) \qquad \text{for all } x \in B \ .$$

Then we have

$$y'(x) = y(x+1) - y(x) = u\left(v(x+1)\right) - u\left(v(x)\right)$$

$$= u\left[v(x+1) - v(x) + v(x)\right] - u\left(v(x)\right)$$

$$= u\left[v'(x) + v(x)\right] - u\left(v(x)\right) \ .$$

Next the increasing property of V implies that

$$v'(x) = v(x+1) - v(x) > 0 \ . \tag{3.7}$$

Thus

$$u\left(v(x)\right) + v'(x) = u\left(v(x) + v'(x)\right) - u\left[v(x) + v'(x) - 1\right] + u\left[v(x) + v'(x) - 1\right]$$

$$= u'(v + v' - 1) - u(v + v' - 1)$$

$$= u'(v + v' - 1) - u'(v + v' - 2) + u'(u + v' - 2)$$

$$+ u(v + v' - 1) - u(v + v' - 2) + u(v + v' - 2)$$

$$= u''(v + v' - 2) + 2u'(v + v' - 2) + u(v + v' - 2) .$$

Continuing in this way, we obtain the expression

$$u(v + v') = \sum_{k=0}^{v'(x)} C_{v'(x)}^k u^{(k)}(v) .$$

Finally we have the following formula

$$u\left(v(x)\right)' = \sum_{k=1}^{v'(x)} C_{v'}^k u^{(k)}\left(v(x)\right) \tag{3.8}$$

where $v'(x) \in \mathbb{N}$ and $u^{(k)}\left(v(x)\right) = u^{(k)}(v)$. In this stage some examples will be useful.

Example 3. Derivative of a constant function is equal to zero, i.e., if $y = f(x) = c$, then

$$y'(x) = f(x + 1) - f(x) = c - c = 0 .$$

Example 4. Derivative of the identity function is equal to 1, i.e., if $y = f(x) = x$ then

$$y' = f(x + 1) - f(x) = (x + 1) - x = 1 .$$

Example 5. *Derivative of the square function.* Let $y = f(x) = x^2$, then

$$y' = f(x + 1) - f(x) = (x + 1)^2 - x^2 = 2x + 1 .$$

Example 6. *Derivative of the power function.* Let $y = f(x) = x^k$, then

$$y'(x) = f(x+1) - f(x) = (x+1)^k - x^k = \sum_{m=0}^{k} C_k^m x^m - x^k$$

that is,

$$\left(x^k\right)' = \sum_{m=0}^{k-1} C_k^m x^m \ . \tag{3.9}$$

Example 7. *Derivative of the arrangement power function.* As we have seen in Section 1.7, the arrangement power function is given by

$$f(x) = x^{[k]} = x(x-1)\cdots[x-(k-1)]$$

where $k \in \mathbb{N}$ and $x \in \mathbb{Z}$.

By definition we have

$$\begin{aligned}
y' &= f(x+1) - f(x) = (x+1)^{[k]} - x^{[k]} \\
&= (x+1)x\ldots[x+1-(k-1)] - x(x-1)\cdots[x-(k-1)] \\
&= x(x-1)\cdots[x(k-2)]\{x+1-[x-(k-1)]\} \\
&= kx(x-1)\cdots[x-(k-2)] \\
&= kx^{[k-1]}
\end{aligned}$$

that is

$$\left(x^{[k]}\right)' = kx^{[k-1]} \ . \tag{3.10_1}$$

We observe that this formula is the analog of the continuous derivative formula for standard power function $f(x) = x^k$.

Example 8. *Derivative of the arrangement function of negative power.* Let $f(x) = x^{[-k]}$, where $k \in \mathbb{N}$. In Section 1.7 we have seen that

$$f(x) = x^{[-k]} = \frac{1}{(x+1)(x+2)\cdots(x+k)} \ .$$

Then

$$f'(x) = \frac{1}{(x+2)\cdots(x+k+1)} - \frac{1}{(x+1)\cdots(x+k)}$$

$$= \frac{(x+1)-(x+k+1)}{(x+1)(x+2)\cdots(x+k)(x+k+1)} = -kx^{[-k-1]}$$

that is

$$\left(x^{[-k]}\right)' = -kx^{[-k-1]} \qquad (3.10_2)$$

which is again similar to the corresponding formula in differential calculus. In next section, we will use both (3.10_1) and (3.10_2) to determine the sum of many interesting finite and infinite series as well.

Example 9. *Derivative of the sin function.* If $y = \sin x$

$$y' = \sin(x+1) - \sin x = 2\sin\frac{1}{2}\cos\left(x+\frac{1}{2}\right)$$

that is,

$$(\sin x)' = 2\sin\frac{1}{2}\cos\left(x+\frac{1}{2}\right)$$

where $x \in \mathbb{Z}$ is given in radians.

Example 10. *Derivative of the logarithm function.* If $y = \ln x$, then

$$y' = \ln(x+1) - \ln x = \ln\left(1+\frac{1}{x}\right)$$

that is,

$$(\ln x)' = \ln\left(1+\frac{1}{x}\right) .$$

Example 11. *Derivative of the exponential function.* If $y = a^x$ where $a > 0$, $a \neq 1$, $a \in \mathbb{R}$, $x \in \mathbb{Z}$. Then

$$y' = a^{x+1} - a^x = a^x(a-1) .$$

Corollary. *Setting $a = 2$ in example 11, we have*

$$(2^x)' = 2^x .$$

In this way, we find a function which is equal to its derivative. This function can be regarded as a discrete analog of standard exponential function in mathematical analysis. It is referred to as the discrete exponential function.

Example 12. *Derivative of the function $y = f(x) = (1+\lambda)^x$, where $\lambda \neq -1$.* We have

$$y' = f(x+1) - f(x) = (1+\lambda)^{x+1} - (1+\lambda)^x = (1+\lambda)^x(1+\lambda-1)$$

$$= (1+\lambda)^x(1+\lambda-1) = \lambda(1+\lambda)^x$$

that is,

$$[(1+\lambda)^x]' = \lambda(1+\lambda)^x .$$

As we see that the function $y = (1+\lambda)^x$ looks like the function $y = e^{\lambda x}$ in analysis. In fact, this is the function which will be used to find a particular solution of a discrete differential equation with constant coefficients. We will return to this fact in the next section.

Having introduced the definition of the derivative of a discrete function, we are now ready to examine the concept of Taylor expansion. To this end, suppose that $f : \mathbb{Z} \to \mathbb{R}$ be any function. Let $x_0 \in \mathbb{Z}$ is a certain number, we wish to express the function f as a series expansion of the form

$$f(x) = \sum_{k=0}^{\infty} a_k \frac{(x-x_0)^{[k]}}{k!} \qquad x \geq x_0$$

where a_k $(k = 0, 1, \dots)$ is a real sequence which has to be found. To find a_0 we replace x by x_0 in the given expression and get $a_0 = f(x_0)$. Now differentiating both sides of the expression we get

$$f'(x) = \sum_{k=1}^{\infty} a_k \frac{(x-x_0)^{[k-1]}}{(k-1)!} = \sum_{k=0}^{\infty} a_{k+1} \frac{(x-x_0)^{[k]}}{k!} .$$

Again replacing x by x_0 we get $a_1 = f'(x_0)$. Continuing in this way, for each $k \geq 0$ we obtain $a_k = f^{(k)}(x_0)$. Therefore,

$$f(x) = \sum_{k=0}^{\infty} \frac{f^{(k)}(x_0)}{k!}(x - x_0)^{[k]} \quad x \geq x_0 \ .$$

Remark. According to the definition of the function $x^{[k]}$, for each $k > x - x_0$, we have $(x - x_0)^{[k]} = 0$. Thus the above expansion can be reduced to the form

$$f(x) = \sum_{k=0}^{x-x_0} \frac{f^{(k)}(x_0)}{k!}(x - x_0)^{[k]} \ .$$

As we see that the above expansion is a finite sum which become larger by increasing the value of x.

To better understand this discrete expansion we give a couple of examples.

Example 12. We have already observed the exponential function $y = 2^x$. Here, we show that its Taylor series is also the same as the Taylor series of the standard exponential function $y = e^x$. In fact, we have

$$2^x = \sum_{k=0}^{x} \frac{x^{[k]}}{k!} \ .$$

We stop here to discuss a geometric-combinatorics problem. Let $f(x)$ denote the number of regions in which x hyperspaces divide the n–dimensional space $\mathbb{R}^n$ provided that these hyperspaces are in general position, i.e., no two hyperspaces are parallel and no three hyperspaces intersect at a single $(n-2)$–affine subspace, no four hyperspaces intersect at a single $(n-3)$–affine subspaces and so on. For each fixed n, one can easily show by induction on x that

$$f(x) = \sum_{k=0}^{n} \frac{x^{[k]}}{k!} \ .$$

Now setting $n = x$, that is the number of hyperspaces is equal to the dimension of the space involved, we get

$$f(x) = \sum_{k=0}^{x} \frac{x^{[k]}}{k!} = 2^x \ ,$$

which gives us the discrete exponential function $f(x) = 2^x$.

Example 13. The discrete exponential function $f(x) = (1 + \lambda)^x$ immediately suggests the following definition.

Let us define,

$$\sin[x] = \frac{(1+i)^x - (1-i)^x}{2i} \ , \quad \cos[x] = \frac{(1+i)^x + (1-i)^x}{2}$$

where $i^2 = -1$.

It can be easily show that

$$\cos[x] = \sum_{k=0}^{\infty} (-1)^k \frac{x^{[2k]}}{(2k)!} = 1 - \frac{x^{[2]}}{2!} + \frac{x^{[4]}}{4!} + \cdots$$

and

$$\sin[x] = \sum_{k=0}^{\infty} (-1)^k \frac{x^{[2k+1]}}{(2k+1)!} = x - \frac{x^{[3]}}{3!} + \frac{x^{[5]}}{5!} + \cdots \ .$$

Also, we have $(\sin[x])' = \cos[x]$, $(\cos[x])' = -\sin[x]$, and

$$(1+i)^x = \cos[x] + i\sin[x] \ ,$$

where the last expression is the analog of Euler's formula. It is referred to as the discrete Euler formula. Later we will use this formula as well as the discrete exponential function $f(x) = 2^x$ to define the discrete complex exponential function which turned out to be an analytic function in discrete concept.

We conclude this section with one last example which gives us an identity analog to the Newton binomial formula.

Example 14. For $a, b \in \mathbb{Z}$ and $n \in \mathbb{N}$ such that $n \geq b$

$$(a+b)^{[n]} = \sum_{k=0}^{n} \binom{n}{k} a^{[k]} b^{[n-k]} . \qquad \text{(Vandermonde's formula)}$$

There are several ways to prove this identity, but our objective here is to use the Taylor series, that is $f(x) = x^{[n]}$ then for $x \geq a$, we have

$$f(x) = \sum_{k=0}^{\infty} \frac{f^{(k)}(a)}{k!} (x-a)^{[k]} .$$

Since $f^{(k)}(a) = n(n-1)\cdots(n-(k-1))a^{[n-k]}$ we get

$$(a+b)^{[n]} = f(a+b) = \sum_{k=0}^{\infty} \binom{n}{k} a^{[n-k]} b^{[k]}$$

$$= \sum_{k=0}^{n} \binom{n}{k} a^{[n-k]} b^{[k]}$$

which is the required identity.

3.2   Discrete integral and its properties

As we have seen in previous section, the definition of the discrete derivative, roughly speaking, is analogous to the continuous one. In present section, we are going to examine the meaning of the so-called discrete integral which even though is not more than a usual sum, it behaves like an ordinary integral not only from its algebraic viewpoint but also the way which it is related to the discrete derivative.

Suppose that f and g are two discrete functions such that $f'(t) = g(t)$ where, $t \in \mathbb{Z}$. If $x_0 \in \mathbb{Z}$ is a fixed number and $x \in \mathbb{Z}$ is an arbitrary point such that $x \geq x_0$, then according to the definition of the discrete derivative we have

$$f(x_0 + 1) - f(x_0) = f'(x_0) = g(x_0)$$

$$f(x_0 + 2) - f(x_0 + 1) = f'(x_0 + 1) = g(x_0 + 1)$$

$$\vdots$$

$$f(x - 1) - f(x - 2) = f'(x - 2) = g(x - 2)$$

$$f(x) - f(x - 1) = f'(x - 1) = g(x - 1) .$$

Summing both sides we get

$$f(x) - f(x_0) = \sum_{t=x_0}^{x-1} g(t)$$

or equivalently

$$f'(x) = f(x_0) + \sum_{t=x_0}^{x-1} g(t) . \tag{3.11}$$

Having said that, the following definition seems to be natural.

Definition of the discrete integral.   Let $f(t)$, $t \in A \subseteq \mathbb{Z}$ be a discrete function, then the expression $\sum_{t=x_0}^{x-1} f(t)$ is called the discrete integral of f on $[x_0, x]$ and is denoted by

$$\int_{x_0}^{x} f(t) \Delta t$$

where $\Delta t = 1$, simply denotes the summation increment.

Using this definition, one can express (3.11) as

$$\int_{x_0}^{x} g(t) \Delta t = f(x) - f(x_0) \tag{3.12}$$

where $g(t) = f'(t)$.

Remark.   Note that $(f(t) + c)' = g(t)$.

This formula looks like the definite integral in calculus, and it can be regarded as the fundamental theorem in difference calculus.

Some Elementary Properties

Property I.   The integral of algebraic sum is equal to the sum of their integral, that is

$$\int_{x_0}^{x} (f(t) + g(t))\, \Delta t = \int_{x_0}^{x} f(t)\Delta t + \int_{x_0}^{x} g(t)\Delta t \ .$$

In fact, this is an immediate consequence of sigma property and it can easily be extended to any finite number of functions or any countable number if the order of the functions involved may be altered without affecting its sum.

Property II.   If the lower limit of the integral sign is equal to the upper limit, then the integral is equal to zero, that is,

$$\int_{x_0}^{x_0} f(t)\Delta t = 0 \ .$$

Property III.   If c is a fixed number, then

$$\int_{x_0}^{x} cf(t)\Delta t = c\int_{x_0}^{x} f(t)\Delta t$$

so that $\int$ and c may interchange order without affecting the result of integration.

Property IV. *Integration by parts formula.*   If $u(t)$ and $v(t)$ are two discrete functions, then

$$\int_{x_0}^{x} u'(t)v'(t)\Delta t = u(t)v(t)\Big|_{x_0}^{x} - \int_{x_0}^{x} u'(t)v(t)\Delta t - \int_{x_0}^{x} u(t)v'(t)\Delta t \ . \quad (3.13)$$

From (3.13) we have the following formula which can be regarded as the integration by parts

$$\int_{x_0}^{x} u'(t)v(t)\Delta t = u(x)v(x) - u(x_0)v(x_0) - \int_{x_0}^{x} u(t)v'(t)\Delta t$$
$$- \int_{x_0}^{x} u'(t)v'(t)\Delta t \ . \quad (3.14)$$

We are going now to examine some elementary but useful formulas.

Example 16. *Arithmetic progression.* From (3.2) and (3.3) we have

$$a'(x) = d \quad \text{and} \quad a(1) = a_1$$

then

$$\int_1^x a'(x)\Delta x = \int_1^x d\Delta x$$

or, $a(x) \big|_1^x = dx \big|_1^x$ which implies that

$$a(x) - a(1) = d(x - 1)$$

or

$$a(x) = a_1 + d(x - 1) \; .$$

Example 17. *The integral of the arrangement power function.* We saw in Example 7 that $\left(x^{[k]}\right)' = kx^{[k-1]}$. Therefore

$$\int_{x_0}^x t^{[k]}\Delta t = \int_{x_0}^x \frac{\left(t^{[k+1]}\right)'}{(k+1)}\Delta t = \frac{t^{[k+1]}}{k+1}\bigg|_{x_0}^x$$

$$\frac{x^{[k+1]}}{k+1} - \frac{x_0^{[k+1]}}{k+1} \qquad x \geq x_0 \; .$$

This integration formula can be used to derive a general summation formula. For this purpose we define, Stirling numbers (named for their discoverer) as follows:

$$x^{[0]} = x^0 = s(0,0) = S(0,0) = 1$$

$$x^{[n]} = x(x-1)\cdots(x-n+1) = \sum_{k=0}^n s(n,k)x^k \qquad n > 0$$

$$x^n = \sum_{k=0}^n S(n,k)x^{[k]} \qquad n > 0 \; .$$

The coefficients $s(n,k)$ and $S(n,k)$ are called Stirling numbers of the first and second kinds respectively. Note that both kinds of numbers are non-zero for $k = 1, 2, \ldots, n$, and for given n, or given k, the numbers of the first kind $s(n,k)$ alternate in sign; in fact, since

$$(-x)^{[n]} = (-1)^n x(x+1) \cdots (x+n-1)$$

it follows from the above formula for $x^{[n]}$ that $(-1)^{n+k} s(n,k)$ is always positive.

Recurrence relations for these numbers follows from

$$x^{[n+1]} = x^{[n]}(x-n) ,$$

in fact we have

$$s(n+1, k) = s(n, k-1) - ns(n, k)$$

and

$$S(n+1, k) = S(n, k-1) + kS(n, k) .$$

It can be easily shown that

$$S(n, k) = \frac{1}{k!} \sum_{j=0}^{k} (-1)^{k-j} \binom{k}{j} j^n .$$

We use the last expression to calculate the finite sum

$$A(m, n) = 1^n + 2^n + \cdots + m^n \qquad n \geq 1 .$$

We have

$$
\begin{aligned}
A(m,n) &= \sum_{x=1}^{m} x^n = \sum_{x=1}^{m} \left(\sum_{k=0}^{n} S(n,k) x^{[k]} \right) \\
&= \sum_{k=0}^{n} S(n,k) \left(\sum_{x=1}^{m} x^{[k]} \right) \\
&= \sum_{k=0}^{n} S(n,k) \left(\int_{1}^{m+1} x^{[k]} \Delta x \right) \\
&= \sum_{k=0}^{n} S(n,k) \left(\frac{x^{[k+1]}}{k+1} \bigg|_{1}^{m+1} \right) \\
&= \sum_{k=0}^{n} S(n,k) \frac{(m+1)^{[k+1]}}{k+1} \\
&= \sum_{k=0}^{n} S(n,k) \frac{m+1}{k+1} \frac{m!}{(m-k)!} \ .
\end{aligned}
$$

Example 18. *Integral of the negative arrangement function.* Since

$$
f(x) = x^{[-k]} = \frac{1}{(x+1)(x+2)\cdots(x+k)} \ ,
$$

and $f'(x) = -k x^{[-k-1]}$, we have

$$
\int_{x_0}^{x} t^{[-k]} \Delta t = \frac{t^{[-k+1]}}{-k+1} \bigg|_{x_0}^{x} = \frac{x^{[1-k]}}{1-k} - \frac{x_0^{[1-k]}}{1-k} \quad k \geq 2 \ .
$$

This integration can be used to calculate the following finite and infinite sums

(i) $\sum_{x=0}^{n} \frac{1}{(x+1)(x+2)\cdots(x+k)}$ $k \geq 2,$

(ii) $\sum_{x=0}^{\infty} \frac{1}{(x+1)(x+2)\cdots(x+k)}$ $k \geq 2.$

In fact, we have

(i) $\sum\limits_{x=0}^{n} \frac{1}{(x+1)(x+2)\cdots(x+k)}$ $= \int_{x=0}^{n+1} x^{[-k]} \Delta x = \frac{x^{[-k+1]}}{-k+1} \Big|_0^{n+1}$

$$= \frac{1}{(1-k)(x+1)(x+2)\cdots(x+k-1)} \Big|_0^{n+1}$$

$$= \frac{1}{1-k} \left[\frac{1}{(n+2)(n+3)\cdots(n+k)} - \frac{1}{(k-1)!} \right].$$

(ii) Letting $n \to \infty$ in (i) we have

$$\sum_{x=0}^{\infty} \frac{1}{(x+1)(x+2)\cdots(x+k)} = \frac{1}{(k-1)(k-1)!}.$$

Example 19. *Integral of the exponential function 2^x.*

$$\int_{x_0}^{x} 2^t \Delta t = 2^t \Big|_{x_0}^{x} = 2^x - 2^{x_0}$$

setting $x_0 = 0$ and $x = n$ we have

$$2^0 + 2^1 + 2^2 + \cdots + 2^{n-1} = \sum_{t=0}^{n-1} 2^t = \int_{t=0}^{n} 2^t \Delta t = 2^n - 2^0$$

or equivalently,

$$2^1 + 2^2 + \cdots + 2^{n-1} = 2^n - 2.$$

We conclude this section with two last examples which are, as a matter of fact, applications of the integral by parts formula.

Example 20.   Evaluate the finite sum

$$\sum_{k=1}^{n} k(k-1)2^k.$$

We have

$$\sum_{k=1}^{n} k(k-1)2^k = \int_{1}^{n+1} x^{[2]} 2^x \Delta x = 2^t t^{[2]} \Big|_{1}^{n+1} - 4 \int_{1}^{n+1} t 2^t \Delta t$$

but

$$\int_1^{n+1} t2^t = t2^t \Big|_1^{n+1} - 2\int_1^{n+1} 2^t = (n+1)2^{n+1} - 2 - 2\,2^t \Big|_1^{n+1}$$

$$= (n+1)2^{n+1} - 2 + 2(2 - 2^{n+1}) = (n-1)2^{n+1} + 2 .$$

Therefore, we have

$$\sum_{k=1}^n k(k-1)2^k = (n+1)n2^{n+1} - 4\left[2 + (n-1)2^{n+1}\right]$$

$$= 2^{n+1}(n^2 - 3n + 4) - 8 .$$

Exercise 21. Evaluate the finite sum

$$\sum_{k=1}^n n(n-1)\cdots(n-k+1)2^k .$$

Hint: Use the Integral

$$\int_1^{n+1} t^{[k]}2^k \Delta t .$$

Example 22. Evaluate the infinite sum

$$f(n) = \sum_{k=0}^\infty 2^{-k}k(k-1)\cdots(k-n+1) \qquad n \geq 2 .$$

We have

$$f(n) = \sum_{k=0}^\infty k^{[n]}2^{-k} = \int_0^\infty t^{[n]}2^{-t}\Delta t = -t^{[n]}2^{-t+1}\Big|_0^\infty + n\int_0^\infty t^{[n-1]}2^{-t}\Delta t$$

$$= 0 + n\int_0^\infty t^{[n-1]}2^{-t}\Delta t = n!\int_0^\infty 2^{-t}\Delta t = n!\left(-2^{-t+1}\right)\Big|_0^\infty = n!$$

i.e.,

$$n! = \sum_{k=0}^\infty k^{[n]}2^{-k} = \sum_{k=1}^\infty 2^{-k}k(k-1)\cdots(k-n+1) .$$

This can be interpreted as a discrete gamma function and may suggest the following discrete generalization of $n!$ for negative integers.

Example 23. Suppose

$$F(n) = \sum_{k=0}^{\infty} 2^{-k} k^{[n]} \qquad n < 0 .$$

Then

$$F(-m) = \int_{x=0}^{\infty} 2^{-x} x^{[-m]} \Delta x \qquad m > 0 .$$

Integrating by parts gives

$$\int_{x=0}^{\infty} 2^{-x} x^{[-m]} \Delta x = \frac{1}{-m+1} x^{[-m+1]} 2^{-x} \Big|_0^{\infty} + \int_0^{\infty} \frac{1}{-m+1} x^{[-m+1]} 2^{-(x+1)} \Delta x$$

$$+ \int_0^{\infty} x^{[-n]} 2^{-(x+1)} \Delta x$$

which yields the following recurrence relation

$$F(-m) = \frac{2}{(m-1)!} + \frac{1}{(-m+1)} F(-m+1) . \qquad (*)$$

Before solving this equation we note that

$$\ln x = \sum_{k=1}^{\infty} \frac{1}{k} \left(\frac{x-1}{x} \right)^k \qquad x \geq \frac{1}{2}$$

hence

$$\sum_{k=1}^{\infty} \frac{1}{2^k k} = \ln 2 .$$

Using this last expression we observe that

$$F(-1) = \sum_{k=0}^{\infty} 2^{-k} k^{[-1]} = \sum_{k=0}^{\infty} \frac{1}{2^k (k+1)}$$

$$= \sum_{k=0}^{\infty} \frac{2}{2^{k+1} (k+1)} = 2 \sum_{k=1}^{\infty} \frac{1}{2^k k} = 2 \ln 2 .$$

Now for the solution of $(*)$ we have

$$F(-m) = \begin{cases} \dfrac{2}{(m-1)!} - \dfrac{1}{(m-1)\cdots[m-(2k-1)]}F(-m+2k-1) \\ \dfrac{1}{(m-1)\cdots(m-2k)}F(-m+2k) \end{cases} \qquad k \geq 1$$

Thus for $m = 2k$, we have

$$F(-2k) = \frac{1}{(2k-1)!} - \frac{1}{(2k-1)!}F(-1) = \frac{1}{(2k-1)!}(1 - 2\ln 2) .$$

Similarly for $m = 2k + 1$, we get

$$F(-(2k+1)) = \frac{1}{(2k)!}F(-1) = \frac{2\ln 2}{(2k)!} .$$

3.3 Solution of the Cauchy problem

Consider

$$\sum_{j=0}^{k} a_j y^{(j)}(x) = 0 \qquad x \geq 0 . \tag{3.15}$$

We wish to solve this equation with the initial condition

$$y^{(j)}(0) = y_j \qquad j = 0, 1, \ldots, k-1 . \tag{3.16}$$

First we seek for a particular solution which is assumed to be in the form of

$$y(x) = (1 + \lambda)^x .$$

By substituting this expression in (3.15) we get the following characteristic equation in λ

$$F(\lambda) = \sum_{j=0}^{k} a_j \lambda^j = 0 . \tag{3.17}$$

If the roots of this equation which are represented by λ_p $(p = 1, \ldots, k)$ are all distinct and different from -1, that is

$$\lambda_p \neq \lambda_q \text{ if } p \neq q , \quad \lambda_p \neq -1 \quad (p, q = 1, \ldots, k) .$$

Then the general solution for (3.15) is

$$y(x) = \sum_{p=1}^{k} C_p (1 + \lambda_p)^x \qquad (3.18)$$

where C_p $(p = 1, \ldots, k)$ are constants yet to be found. Considering the initial condition (3.16) in (3.18) we obtain the following linear system of equations for the constants C_p

$$\sum_{p=1}^{k} C_p \lambda_p^j = y_j \qquad j = 0, 1, \ldots, k-1 . \qquad (3.19)$$

Since the determinant of this system is Vandermonde determinant and this is different from zero by the above condition for λ_p's, we deduce that this system has a unique solution by Cramer's rule. Substituting this solutions in (3.18) we obtain the solution for the Cauchy problem. It is clear from the above scheme that if the roots of characteristic equation are distinct and all different from zero, then the solution is determined uniquely.

Example 24. *Solution of the Fibonacci sequence.* As we have seen at the beginning of this chapter, to determine the Fibonacci sequence we have to solve the Cauchy problem

$$\begin{cases} a''(x) + a'(x) - a(x) & = 0 \\ a(0) & = \alpha \qquad x \geq 0 \\ a'(0) & = \beta . \end{cases}$$

The characteristic equation is

$$\lambda^2 + \lambda - 1 = 0 ,$$

which yields

$$\lambda_j = \frac{-1 + (-1)^j \sqrt{5}}{2} \qquad (j = 1, 2) .$$

Therefore from (3.18) we have the general solution as

$$a(x) = \sum_{p=1}^{2} C_p (1 + \lambda_p)^x = \sum_{p=1}^{2} C_p \left(\frac{1 + (-1)^p \sqrt{5}}{2} \right)^x \qquad x \geq 0 .$$

Now by using the initial conditions for the constant C_p $(p = 1, 2)$ we get the following system of equations

$$C_1 + C_2 = a(0) = \alpha$$

$$\frac{-1 - \sqrt{5}}{2} C_1 + \frac{-1 + \sqrt{5}}{2} C_2 = a'(0) = \beta .$$

Accordingly by the Cramer's rule we have

$$\Delta = \begin{vmatrix} 1 & 1 \\ \frac{-1-\sqrt{5}}{2} & \frac{-1+\sqrt{5}}{2} \end{vmatrix} = 5$$

$$\Delta_1 = \begin{vmatrix} \alpha & 1 \\ \beta & \frac{-1+\sqrt{5}}{2} \end{vmatrix} = \frac{-1 + \sqrt{5}}{2} \alpha - \beta$$

$$\Delta_2 = \begin{vmatrix} 1 & \alpha \\ \frac{-1-\sqrt{5}}{2} & \beta \end{vmatrix} = \beta - \frac{-1 - \sqrt{5}}{2} \alpha .$$

Thus

$$C_1 = \frac{\Delta_1}{\Delta} = \frac{(\sqrt{5} - 1)\alpha - 2\beta}{2\sqrt{5}}$$

$$C_2 = \frac{\Delta_2}{\Delta} = \frac{(\sqrt{5} + 1)\alpha + 2\beta}{2\sqrt{5}} .$$

By substituting these values in the above expression we have the general solution as

$$a(x) = \frac{(\sqrt{5} - 1)\alpha - 2\beta}{2\sqrt{5}} \left(\frac{1 - \sqrt{5}}{2} \right)^x + \frac{(\sqrt{5} + 1)\alpha + 2\beta}{2\sqrt{5}} \left(\frac{1 + \sqrt{5}}{2} \right)^x \qquad x \geq 0 .$$

In this way, we solved the Cauchy problem for the Fibonacci sequence. If $\alpha = a_0$ and $\beta = a_1 - a_0$, then we have a Fibonacci sequence in which the first and second terms respectively are a_0 and $a_1 - a_0$.

3.4   Methods of the variation of parameters

Consider the differential equation

$$\sum_{j=0}^{k} a_j y^{(j)}(x) = f(x) \qquad x \geq 0 . \tag{3.20}$$

Assuming $f(x) = 0$ in this equation will lead to the equation (3.15). In other words (3.15) is the corresponding homogeneous equation of (3.20). If the roots of the characteristic equation i.e., λ_p $(p = 1, \ldots, k)$ are all distinct and different from (-1), then the homogeneous equation has a solution in the form of (3.18). Now we are going to examine the general solution of (3.20) as follows

$$y(x) = \sum_{p=1}^{k} C_p(x)(1 + \lambda_p)^x \tag{3.21}$$

where $C_p(x)$ are arbitrary functions. By discrete differentiating of both sides of (3.21) with respect to x we obtain

$$y'(x) = \sum_{p=1}^{k} C_p(x)\lambda_p(1 + \lambda_p)^x + \sum_{p=1}^{k} C_p'(x)(1 + \lambda_p)^{x+1} .$$

Assuming

$$\sum_{p=1}^{k} C_p'(x)(1 + \lambda_p)^{x+1} = 0 \tag{3.22$_1$}$$

the expression for $y'(x)$ gives rise to

$$y'(x) = \sum_{p=1}^{k} C_p(x)\lambda_p(1 + \lambda_p)^x . \tag{3.21$_1$}$$

Similarly by differentiating $y'(x)$ in (3.21_1) we have

$$y''(x) = \sum_{p=1}^{k} C_p(x)\lambda_p^2(1+\lambda_p)^x + \sum_{p=1}^{k} C_p'(x)\lambda_p(1+\lambda_p)^{x+1}$$

from which by accepting

$$\sum_{p=1}^{k} C_p'(x)\lambda_p(1+\lambda_p)^{x+1} = 0 . \tag{3.22_2}$$

We obtain the following expression for $y''(x)$

$$y''(x) = \sum_{p=1}^{k} C_p(x)\lambda_p^2(1+\lambda_p)^x . \tag{3.21_2}$$

By continuing in this procedure up to the $(k-1)^{\text{th}}$ order we have

$$y^{(k-1)}(x) = \sum_{p=1}^{k} C_p(x)\lambda_p^{k-1}(1+\lambda_p)^x + \sum_{p=1}^{k} C_p'\lambda_p^{k-2}(1+\lambda_p)^{x+1}$$

on which by assuming

$$\sum_{p=1}^{k} C_p'(x)\lambda_p^{k-2}(1+\lambda_p)^{x+1} = 0 \tag{3.22_{k-1}}$$

we obtain the expression

$$y^{(k-1)}(x) = \sum_{p=1}^{k} C_p(x)\lambda_p^{k-1}(1+\lambda_p)^x . \tag{3.21_{k-1}}$$

By differentiating (3.21_{k-1}) once more and substituting the result along with the expressions $(3.21), (3.21_1), \ldots, (3.21_{k-1})$ in (3.20) we have the following equation

$$\sum_{j=0}^{k} a_j \sum_{p=1}^{k} C_p(x)\lambda_p^j(1+\lambda_p)^x + a_k \sum_{p=1}^{k} C_p'(x)\lambda_p^{k-1}(1+\lambda_p)^{x+1} = f(x)$$

or equivalently,

$$\sum_{p=1}^{k} C_p(x)(1+\lambda_p)^x \sum_{j=0}^{k} a_j \lambda_p^j + a_k \sum_{p=1}^{k} C_p'(x)\lambda_p^{k-1}(1+\lambda_p)^{x+1} = f(x)$$

where the first term is equal to zero. Since $a_k \neq 0$, dividing both sides by a_k we get

$$\sum_{p=1}^{k} C_p'(x)\lambda_p^{k-1}(1+\lambda_p)^{x+1} = f(x)/a_k . \qquad (3.22_k)$$

The expressions $(3.22_1), (3.22_2), \ldots, (3.22_k)$ give the following system of equations for $C_p'(x)$ $(p = 1, \ldots, k)$

$$\begin{cases} \sum_{p=1}^{k} C_p'(x)\lambda_p^{s}(1+\lambda_p)^{x+1} = 0 , \quad s = 0,1,\ldots,k-2 \\ \\ \sum_{p=1}^{k} C_p'(x)\lambda_p^{k-1}(1+\lambda_p)^{x+1} = f(x)/a_k . \end{cases} \qquad (3.22)$$

Since the determinant of this system, that is

$$\Delta(x) = \begin{vmatrix} (1+\lambda_1)^{x+1} & (1+\lambda_2)^{x+1} & \cdots & (1+\lambda_k)^{x+1} \\ \lambda_1(1+\lambda_1)^{x+1} & \lambda_2(1+\lambda_2)^{x+1} & \cdots & \lambda_k(1+\lambda_k)^{x+1} \\ \vdots & \vdots & & \vdots \\ \lambda_1^{k-1}(1+\lambda_1)^{x+1} & \lambda_2^{k-1}(1+\lambda_2)^{x+1} & \cdots & \lambda_k^{k-1}(1+\lambda_k)^{x+1} \end{vmatrix}$$

is different from zero, by Cramer's rule we obtain the following solution

$$
C'_p(x) = \frac{1}{\Delta x}
\begin{vmatrix}
1 & 1 & \cdots & 1 & 0 & \cdots & 1 \\
\lambda_1 & \lambda_2 & \cdots & \lambda_{p-1} & 0 & \cdots & \lambda_k \\
\vdots & \vdots & & \vdots & \vdots & & \vdots \\
\lambda_1^{k-1} & \lambda_2^{k-1} & \cdots & \lambda_{p-1}^{k-1} & \frac{f(x)}{a_k} & \cdots & \lambda_k^{k-1}
\end{vmatrix}
$$

$$
= \frac{(1+\lambda_p)^{-x-1}}{D(\lambda_1,\ldots,\lambda_k)}(-1)^{k+p}\frac{f(x)}{a_k}D(\lambda_1,\ldots,\lambda_k)
$$

$$
= \frac{(1+\lambda_p)^{-x-1}}{\prod_{j\neq p}(\lambda_p-\lambda_j)}\frac{f(x)}{a_k} \quad (p=1,\ldots,k) .
$$

By discrete integration we get

$$
C_p(x) - C_p(0) = \sum_{\xi=0}^{x-1} \frac{(1+\lambda_p)^{-\xi-1}}{\prod_{j\neq p}(\lambda_p-\lambda_j)}\frac{f(\xi)}{a_k}
$$

$$
= \int_0^x \frac{(1+\lambda_p)^{-\xi-1}}{\prod_{j\neq p}(\lambda_p-\lambda_j)}\frac{f(\xi)}{a_k}\Delta\xi
$$

or equivalently

$$
C_p(x) = C_p(0) + \int_0^x \frac{(1+\lambda_p)^{-\xi-1}}{\prod_{j\neq p}(\lambda_p-\lambda_j)}\frac{f(\xi)}{a_k}\Delta\xi \quad (p=1,\ldots,k) .
$$

By substituting this expression in (3.21) we get the following general solution for (3.20)

$$
y(x) = \sum_{p=1}^{k} C_p(1+\lambda_p)^x + \int_0^x \frac{1}{a_k}\sum_{p=1}^{k}\frac{(1+\lambda_p)^{x-\xi-1}}{\prod_{j\neq p}(\lambda_p-\lambda_j)}f(\xi) . \tag{3.23}
$$

By letting

$$
g(x-\xi) = \frac{1}{a_k}\sum_{p=1}^{k}\frac{(1+\lambda_p)^{x-\xi-1}}{\prod_{j\neq p}(\lambda_p-\lambda_j)} \quad 0\leq\xi<x . \tag{3.24}
$$

the above solution can be rewritten in the following closed form

$$y(x) = \sum_{p=1}^{k} C_p (1 + \lambda_p)^x + \int_0^x g(x - \xi) f(\xi) \Delta\xi$$

where the function in (3.24) is called the fundamental solution of (3.20).

3.5   Solution of Cauchy problem for non-homogeneous differential equation

We first examine some properties of non-homogeneous differential equation. As we have seen that the general solution of (3.20) is in the form of (3.23_1) where the first term is the general solution of (3.15) and the second term is a particular solution of (3.20). In fact, to prove this we observe that since

$$y_0(x) = \int_0^x g(x - \xi) f(\xi) \Delta\xi$$

then the value of $y_0(x)$ is

$$y_0(x) = \frac{1}{a_k} \sum_{\xi=0}^{x-1} f(\xi) \sum_{p=1}^{k} (1 + \lambda_p)^{x-\xi-1} \frac{D_p(\lambda_1, \ldots, \lambda_k)}{D(\lambda_1, \ldots, \lambda_k)} \qquad (3.25)$$

where

$$D_p(\lambda_1, \ldots, \lambda_k) = (-1)^{k+p} D(\lambda_1, \ldots, \lambda_{p-1}, \lambda_{p+1}, \ldots, \lambda_k)$$

is the corresponding cofactor of $(k, p)^{\text{th}}$ entry in Vandermonde determinant.

Now by one of the properties of the discrete integral we have

$$y_0(0) = 0 \ .$$

Taking the derivative of (3.25) we get

$$y_0'(x) = \frac{1}{a_k} \sum_{\xi=0}^{x} f(\xi) \sum_{p=1}^{k} (1+\lambda_p)(1+\lambda_p)^{x-\xi-1}\frac{D_p}{D}$$

$$- \frac{1}{a_k} \sum_{\xi=0}^{x-1} f(\xi) \sum_{p=1}^{k} (1+\lambda_p)^{x-\xi-1}\frac{D_p}{D}$$

$$= \frac{1}{a_k} f(x) \sum_{p=1}^{k} \frac{D_p}{D} + \frac{1}{a_k} \sum_{\xi=0}^{x-1} f(\xi) \sum_{p=1}^{k} \lambda_p(1+\lambda_p)^{x-\xi-1}\frac{D_p}{D}$$

where the first term is equal to zero, because $\sum_{p=1}^{k} D_p = 0$. Thus

$$y_0'(x) = \frac{1}{a_k} \sum_{\xi=0}^{x-1} f(\xi) \sum_{p=1}^{k} \lambda_p(1+\lambda_p)^{x-\xi-1}\frac{D_p}{D}$$

$$= \int_0^x g'(x-\xi)f(\xi)\Delta\xi \ .$$

(3.26)

Continuing in this procedure and taking into account the above property of the determinant at each stage we observe that

$$y_0^{(j)}(x) = \int_0^x g^{(j)}(x-\xi)f(\xi) \qquad j = 0, 1, \ldots, k-1$$

(3.26₁)

where

$$y_0^{(j)}(0) = 0 \qquad j = 0, 1, \ldots, k-1 \ .$$

(3.27)

Eventually if we differentiate the expression (3.26_1) for $j = k - 1$, then

$$y_0^{(k)}(x) = \left(\frac{1}{a_k} \sum_{\xi=0}^{x-1} f(\xi) \sum_{p=1}^{k} \lambda_p^{k-1}(1 + \lambda_p)^{x-\xi-1} \frac{D_p}{D} \right)'$$

$$= \frac{1}{a_k} \sum_{\xi=0}^{x} f(\xi) \sum_{p=1}^{k} \lambda_p^{k-1}(1 + \lambda_p)(1 + \lambda_p)^{x-\xi-1} \frac{D_p}{D}$$

$$- \frac{1}{a_k} \sum_{\xi=0}^{x-1} f(\xi) \sum_{p=1}^{k} \lambda_p^{k-1}(1 + \lambda_p)^{x-\xi-1} \frac{D_p}{D}$$

$$= \frac{1}{a_k} f(x) \frac{1}{D} \sum_{p=1}^{k} \lambda_p^{k-1} D_p + \frac{1}{a_k} \sum_{\xi=0}^{x-1} f(\xi) \sum_{p=1}^{k} \lambda_p^{k}(1 + \lambda_p)^{x-\xi-1} \frac{D_p}{D}$$

since λ_p^{k-1} $(p = 1, \ldots, k)$ is the last row in Vandermonde determinant we have

$$\sum_{p=1}^{k} \lambda_p^{k-1} D_p = D .$$

This implies the following result:

$$y_0^{(k)}(x) = \frac{f(x)}{a_k} \sum_{\xi=0}^{x-1} f(\xi) \sum_{p=1}^{k} \lambda_p^{k}(1 + \lambda_p)^{x-\xi-1} \frac{D_p}{D}$$

$$= \frac{1}{a_k} f(x) + \int_0^x g^{(k)}(x - \xi) f(\xi) . \tag{3.26_2}$$

From the expressions (3.26_1) and (3.26_2) we see that

$$g(x - \xi) \qquad \xi \neq x \qquad (0 \leq \xi < x)$$

as a function of independent variable x is the solution of the homogeneous equation of (3.15) corresponding to non-homogeneous equation of (3.20).

We are going now to seek a solution of (3.20) which satisfies the initial condition

$$y^{(j)}(0) = y_j \qquad (j = 0, 1, \ldots, k - 1) . \tag{3.28}$$

For this purpose we determine the coefficients C_p $(p = 1, \ldots, k)$ in expression (3.23_1) under the condition (3.28). Since we have already seen that the solution $y_0(x)$ in (3.35) satisfies the initial condition (3.27), then

$$\sum_{p=1}^{k} C_p \lambda_p^j = y_j \qquad (j = 0, 1, \ldots, k-1) . \qquad (3.29)$$

Since the λ_p's, are all distinct the determinant of the matrix of the coefficients in (3.29) is again the Vandermonde determinant which was already denoted by $D = D(\lambda_1, \ldots, \lambda_n)$. If we denote the cofactor of $(p, q)^{\text{th}}$ entry by D_{pq}, then from (3.29) we have

$$C_p = \frac{1}{D} \sum_{q=1}^{k} y_{q-1} D_{pq} \qquad (p = 1, 2, \ldots, k) . \qquad (3.30)$$

By substituting the expression (3.30) in (3.23_1), for the Cauchy problem (3.20) and (3.28) we obtain the following solution

$$y(x) = \int_0^x g(x - \xi) f(\xi) + \sum_{p=1}^{k} \frac{(1 + \lambda_p)^x}{D} \sum_{q-1}^{k} y_{q-1} D_{pq} .$$

Finally if we denote the $(k+1) \times (k+1)$ determinant

$$D(x) = \begin{vmatrix} 0 & (1 + \lambda_1)^x & \cdots & (1 + \lambda_k)^x \\ y_0 & 1 & \cdots & 1 \\ y_1 & \lambda_1 & \cdots & \lambda_k \\ \vdots & \vdots & & \vdots \\ y_{k-1} & \lambda_1^{k-1} & \cdots & \lambda_k^{k-1} \end{vmatrix} \qquad (3.31)$$

by $D(x)$, then the solution of the Cauchy problem takes the following form

$$y(x) = \int_0^x g(x - \xi) f(\xi) - \frac{D(x)}{D} \qquad (3.32)$$

where the first factor is the solution of non-homogeneous equation with initial condition (3.27) and the second factor is the solution of the homogeneous equation with non-zero initial condition (3.28).

Example 25. *Solution of the arithmetic progression.* We have shown in previous discussions that the solution of the arithmetic progression gives rise to the following Cauchy problem

$$\begin{cases} a'(x) & = d \qquad x \geq 0 \\ a(0) & = a_0 \ . \end{cases}$$

We note that $k = 1$, $f(x) = d$, $y_0 = a_0$, $D = 1$, $\lambda = 0$, so $D(x) = -y_0 = -a_0$, $g(x - \xi) = 1$ from which by substituting these values in (3.32) the solution is obtained as

$$a(x) = a_0 + d \int_0^x 1 = a_0 + d \sum_{\xi=0}^{x-1} 1 = a_0 + dx \ .$$

It is clear that if instead of 0 we start from 1, then the general term is expressed as

$$a(x) = a_1 + d(x - 1) \qquad x \geq 1 \ .$$

Finally by replacing $x = k$, $a_k = a(k)$ we have

$$a_k = a_1 + (k - 1)d \qquad k \geq 1 \ .$$

3.6   Solution of the boundary value problem for homogeneous differential equations of second order

Consider the boundary value problem

$$y''(x) - 2ay'(x) - by(x) = 0 \qquad 0 \leq x \leq n \tag{3.33}$$

$$y(0) = y_0$$

$$\tag{3.34}$$

$$y(n) = y_n \ .$$

To solve this problem, like what we did in Cauchy problem, we first find its general solution. It can be readily seen that the expression

$$y(x) = C_1(1 + \lambda_1)^x + C_2(1 + \lambda_2)^x \tag{3.35}$$

is the general solution of (3.33), where C_1 and C_2 are arbitrary constants and λ_1 and λ_2 are the roots of the characteristic equation. For convenience we denote these roots as

$$\lambda_j = a + (-1)^j \sqrt{a^2 + b} \qquad (j = 1, 2) .$$

Now for the constants C_1 and C_2 by using the boundary value condition (3.34) we get the following system of equations

$$\begin{cases} C_1 + C_2 & = y_0 \\ C_1(1 + \lambda_1)^n + C_2(1 + \lambda_2)^n & = y_n . \end{cases} \tag{3.36}$$

If

$$\Delta = \begin{vmatrix} 1 & 1 \\ (1 + \lambda_1)^n & (1 + \lambda_2)^n \end{vmatrix} = (1 + \lambda_2)^n - (1 + \lambda_1)^n$$

$$= (\lambda_2 - \lambda_1) \sum_{p+q=n-1} (1 + \lambda_2)^p (1 + \lambda_1)^q \neq 0 , \tag{3.37}$$

then by Cramer's formula we have

$$C_1 = \frac{1}{\Delta} \begin{vmatrix} y_0 & 1 \\ y_n & (1 + \lambda_2)^n \end{vmatrix} = \frac{y_0(1 + \lambda_2)^n - y_n}{\Delta}$$

$$C_2 = \frac{1}{\Delta} \begin{vmatrix} 1 & y_0 \\ (1 + \lambda_1)^n & y_n \end{vmatrix} = \frac{y_n - (1 + \lambda_1)^n y_0}{\Delta} .$$

By substituting these values in (3.35) for the boundary value problem (3.33)–(3.34) we obtain the following solution

$$y(x) = \frac{y_0(1 + \lambda_2)^n - y_n}{\Delta}(1 + \lambda_1)^x + \frac{y_n - (1 + \lambda_1)^x y_0}{\Delta}(1 + \lambda_2)^x$$

which can be further simplified as

$$y_n = \frac{(1 + \lambda_2)^n(1 + \lambda_1)^x - (1 + \lambda_1)^n(1 + \lambda_2)^x}{(1 + \lambda_2)^n - (1 + \lambda_1)^n} y_0 + \frac{(1 + \lambda_2)^x - (1 - \lambda_1)^x}{(1 + \lambda_2)^n - (1 + \lambda_1)^n} y_n .$$

$$\tag{3.38}$$

Using this expression one can easily show that the boundary conditions hold.

Now let us suppose that the condition (3.27) does not hold, that is we have

$$\Delta = (1+\lambda_2)^n - (1+\lambda_1)^n = 0 .$$

$$(3.37_1)$$

In this case the expression (3.38) is absurd, in other words, there is no solution in the form of (3.38). From the condition (3.37_1) we observe that

$$1+\lambda_2 = (1+\lambda_1)\varepsilon_j$$

where ε_j is a root of unity, i.e., $\varepsilon_j = \omega^j$ where $\omega = \cos\frac{2\pi}{n} + i\sin\frac{2\pi}{n}$ and j is a positive integer such that $1 \le j \le n$.

Therefore in order the (3.37_1) holds, there must be a positive integer $j_0 \in [1, n)$ such that the following is true

$$1+\lambda_2 = (1+\lambda_1)\left(\cos\frac{2\pi j_0}{n} + i\sin\frac{2\pi j_0}{n}\right) .$$

$$(3.39)$$

So the system in (3.36) gives rise to

$$\begin{cases} C_1 + C_2 & = y_0 \\ (1+\lambda_1)^n(C_1 + C_2) & = y_n . \end{cases}$$

$$(3.36_1)$$

Note that the system (3.36_1) may have no solution in general. In this case we say that the boundary value problem in question has no solution.

If

$$y_n = (1+\lambda_1)^n y_0$$

then one of the equations is a scalar multiple of the other. It suffices therefore to deal with only one of them. In this case we have

$$C_2 = y_0 - C_1$$

where C_1 is a parameter. Thus the solution of the boundary value problem is

$$y(x) = c\left[(1+\lambda_1)^x - (1+\lambda_2)^x\right] + y_0(1+\lambda_2)^x$$

where c is an arbitrary constant.

Corollary. *If* (3.37_1) *holds, then the boundary value problem* (3.33)–(3.34) *either has no solution or has infinitely many solutions.*

Now instead of the boundary value condition (3.34) we consider

$$\begin{cases} y'(0) & = \alpha \\ y'(n) & = \beta \ . \end{cases} \tag{3.40}$$

If for the boundary value problem (3.33)–(3.40) we substitute the general solution (3.35) of (3.33) in (3.40), we get the following system of equations for C_1, C_2

$$\begin{cases} C_1\lambda_1 + C_2\lambda_2 & = \alpha \\ C_1\lambda_1(1+\lambda_1)^n + C_2\lambda_2(1+\lambda_2)^n & = \beta \ . \end{cases} \tag{3.41}$$

Since

$$\Delta_1 = \begin{vmatrix} \lambda_1 & \lambda_2 \\ \lambda_1(1+\lambda_1)^n & \lambda_2(1+\lambda_2)^n \end{vmatrix} = \lambda_1\lambda_2\Delta$$

is different from zero, all procedures remain valid.

Finally we consider the boundary value problem

$$L_j y = \alpha_{1j}y'(0) + \beta_{1j}y'(n) + \alpha_j y(0) + \beta_j y(n) = \gamma_j \qquad j = 1,2 \tag{3.42}$$

where $\alpha_{1j}, \beta_{1j}, \alpha_j, \beta_j, \gamma_j$ $(j = 1, 2)$ are given constants. By substituting the general solution (3.35) of the equation (3.33) in the boundary condition (3.42), we get the following system

$$\begin{cases} \alpha_{1j}(C_1\lambda_1 + C_2\lambda_2) + \beta_{1j}\left[C_1\lambda_1(1+\lambda_1)^n + C_2\lambda_2(1+\lambda_2)^n\right] + \\ \alpha_j(C_1+C_2) + \beta_j\left[C_1(1+\lambda_1)^n + C_2(1+\lambda_2)^n\right] = \gamma_j \qquad (j=1,2) \ . \end{cases} \tag{3.43}$$

It is clear that for existence of solution we have to have

$$\Delta_2 \neq 0 . \tag{3.44}$$

In this case by Cramer's rule the constant C_1, C_2 are uniquely determined from the system (3.43). Thus the solution of (3.33)–(3.42) are uniquely determined by (3.35). If the condition (3.44) does not hold, then this problem like the first one has either no solution or infinitely many solutions.

Now using the properties of the determinant, we can rewrite the determinant in (3.34) as the following simple form

$$
\begin{aligned}
\Delta_2 &= \lambda_1\lambda_2(1+\lambda_2)^n \begin{vmatrix} \alpha_{11} & \beta_{11} \\ \alpha_{12} & \beta_{12} \end{vmatrix} + \lambda_1 \begin{vmatrix} \alpha_{11} & \alpha_1 \\ \alpha_{12} & \alpha_2 \end{vmatrix} + \lambda_1(1+\lambda_2)^n \begin{vmatrix} \alpha_{11} & \beta_1 \\ \alpha_{12} & \beta_2 \end{vmatrix} \\
&\quad + \lambda_1\lambda_2(1+\lambda_1)^n \begin{vmatrix} \beta_{11} & \alpha_{11} \\ \beta_{12} & \alpha_{12} \end{vmatrix} + \lambda_2 \begin{vmatrix} \alpha_1 & \alpha_{11} \\ \alpha_2 & \alpha_{12} \end{vmatrix} + \lambda_1(1+\lambda_1)^n \begin{vmatrix} \beta_{11} & \alpha_1 \\ \beta_{12} & \alpha_2 \end{vmatrix} \\
&\quad + \lambda_1(1+\lambda_1)^n(1+\lambda_2)^n \begin{vmatrix} \beta_{11} & \beta_1 \\ \beta_{12} & \beta_2 \end{vmatrix} + \lambda_2(1+\lambda_2)^n \begin{vmatrix} \alpha_1 & \beta_{11} \\ \alpha_2 & \beta_{12} \end{vmatrix} \\
&\quad + \lambda_2(1+\lambda_1)^n(1+\lambda_2)^n \begin{vmatrix} \beta_1 & \beta_{11} \\ \beta_2 & \beta_{12} \end{vmatrix} + \lambda_2(1+\lambda_1)^n \begin{vmatrix} \beta_1 & \alpha_{11} \\ \beta_2 & \alpha_{12} \end{vmatrix} \\
&\quad + (1+\lambda_2)^n \begin{vmatrix} \alpha_1 & \beta_1 \\ \alpha_2 & \beta_2 \end{vmatrix} + (1+\lambda_1)^n \begin{vmatrix} \beta_1 & \alpha_1 \\ \beta_2 & \alpha_2 \end{vmatrix} \\
&= (1+\lambda_1)^n(1+\lambda_2)^n \begin{vmatrix} \beta_1 & \beta_{11} \\ \beta_2 & \beta_{12} \end{vmatrix} (\lambda_2 - \lambda_1) + \lambda_1\lambda_2 \begin{vmatrix} \alpha_{11} & \beta_{11} \\ \beta_2 & \beta_{12} \end{vmatrix} \\
&\quad \times [(1+\lambda_2)^n - (1+\lambda_1)^n] + \begin{vmatrix} \alpha_{11} & \beta_1 \\ \alpha_{12} & \beta_2 \end{vmatrix} [\lambda_1(1+\lambda_2)^n - \lambda_2(1+\lambda_1)^n] \\
&\quad + \begin{vmatrix} \alpha_1 & \beta_{11} \\ \alpha_2 & \beta_{12} \end{vmatrix} [\lambda_2(1+\lambda_2)^n - \lambda_1(1+\lambda_1)^n] \\
&\quad + \begin{vmatrix} \alpha_1 & \alpha_{11} \\ \alpha_2 & \alpha_{12} \end{vmatrix} (\lambda_2 - \lambda_1) + \begin{vmatrix} \alpha_1 & \beta_1 \\ \alpha_2 & \beta_2 \end{vmatrix} [(1+\lambda_2)^n - (1-\lambda_1)^n] .
\end{aligned}
$$

We note that if $\lambda_1 \neq \lambda_2$ and the left-hand sides of (3.42) are not linearly dependent, then the condition $\Delta_2 = 0$ between λ_1 and λ_2 already shows a mixed linear dependence .

3.7   Solution of the boundary value problem for non-homogeneous differential equations of second order

In this section we are going to examine the differential equation

$$y''(x) - 2ay'(x) - by(x) = f(x) \qquad 0 \le x \le n \tag{3.45}$$

with the boundary condition

$$y(0) = y(n) = 0 . \tag{3.46}$$

For this purpose, we first find the general solution of (3.45). By repeating the same procedures used in the solution of Cauchy problem for the general solution of (3.45) we obtain

$$y(x) = \sum_{p=1}^{2} C_{p1}(1 + \lambda_p)^x + \int_0^x g_1(x - \xi)f(\xi)\Delta\xi \tag{3.47}$$

where C_{21} and C_{11} are arbitrary constants and λ_1, λ_2 are roots of the characteristic equation

$$\lambda^2 - 2a\lambda - b = 0$$

where $\lambda_1 \ne \lambda_2$.

Now from the expression (3.24) we may calculate the function $g_1(x - \xi)$ as follows

$$g_1(x - \xi) = \sum_{p=1}^{2} \frac{(1 + \lambda_p)^{x - \xi - 1}}{\prod_{\substack{j \ne p}}^{2}(\lambda_p - \lambda_j)}$$

$$= \frac{(1 + \lambda_1)^{x - \xi - 1}}{\lambda_1 - \lambda_2} + \frac{(1 + \lambda_2)^{x - \xi - 1}}{\lambda_2 - \lambda_1}$$

which is in turn equal to

$$g_1(x - \xi) = \frac{(1 + \lambda_2)^{x-\xi-1} - (1 + \lambda_1)^{x-\xi-1}}{\lambda_2 - \lambda_1} \qquad 0 \le \xi < x .$$

To find the constants appeared in (3.47) from (3.46) we obtain

$$\begin{cases} C_{11} + C_{21} & = 0 \\[2em] C_{11}(1 + \lambda_1)^n + C_{21}(1 + \lambda_2)^n & = -\displaystyle\int_0^n g_1(n - \xi) f(\xi) . \end{cases}$$

If in this system we have

$$\Delta = (1 + \lambda_2)^n - (1 + \lambda_1)^n \neq 0$$

then by Cramer's rule we get the following solutions

$$C_{11} = \frac{1}{\Delta} \int_0^n g_1(n - \xi) f(\xi)$$

$$C_{21} = -\frac{1}{\Delta} \int_0^n g_1(n - \xi) f(\xi) .$$

By substituting these values in (3.47), we obtain the solutions of (3.45)–(3.46) as follows

$$y(x) = \int_0^x g_1(x - \xi) f(\xi) + \frac{(1 + \lambda_1)^x - (1 + \lambda_2)^x}{(1 + \lambda_2)^n - (1 + \lambda_1)^n} \int_0^n g_1(n - \xi) f(\xi) .$$

For simplicity, we rewrite the above expression in the following form. By substituting the expression $g_1(x - \xi)$ and rearranging the similar terms we have

$$y(x) = \sum_{\xi=0}^{x-1} \frac{(1+\lambda_2)^{x-\xi-1} - (1+\lambda_1)^{x-\xi-1}}{\lambda_2 - \lambda_1} \, f(\xi)$$

$$+ \frac{(1+\lambda_1)^x - (1+\lambda_2)^x}{(1+\lambda_2)^n - (1+\lambda_1)^n} \sum_{\xi=0}^{n-1} \frac{(1+\lambda_2)^{n-\xi-1} - (1+\lambda_1)^{n-\xi-1}}{\lambda_2 - \lambda_1} \, f(\xi)$$

$$= \sum_{\xi=0}^{x-1} f(\xi) \frac{(1+\lambda_2)^{n+x-\xi-1} + (1+\lambda_1)^{n-\xi-1}(1+\lambda_2)^x}{(\lambda_2 - \lambda_1)\left[(1+\lambda_2)^n - (1+\lambda_1)^n\right]}$$

$$+ \sum_{\xi=0}^{x-1} f(\xi) \frac{-(1+\lambda_2)^n (1+\lambda_1)^{x-\xi-1} - (1+\lambda_1)^n (1+\lambda_2)^{x-\xi-1}}{(\lambda_2 - \lambda_1)\left[(1+\lambda_2)^n - (1+\lambda_1)^n\right]}$$

$$+ \sum_{\xi=0}^{x-1} f(\xi) \frac{(1+\lambda_1)^{n+x-\xi-1} - (1+\lambda_2)^{n+x-\xi-1}}{(\lambda_2 - \lambda_1)\left[(1+\lambda_2)^n - (1+\lambda_1)^n\right]}$$

$$+ \sum_{\xi=0}^{x-1} f(\xi) \frac{(1+\lambda_1)^x (1+\lambda_2)^{n-\xi-1} - (1+\lambda_1)^{n+x-\xi-1}}{(\lambda_2 - \lambda_1)\left[(1+\lambda_2)^n - (1+\lambda_1)^n\right]}$$

$$+ \sum_{\xi=x}^{n-1} f(\xi) \frac{(1+\lambda_1)^x (1+\lambda_2)^{n-\xi-1} - (1+\lambda_1)^{n+x-\xi-1}}{(\lambda_2 - \lambda_1)\left[(1+\lambda_2)^n - (1+\lambda_1)^n\right]}$$

$$+ \sum_{\xi=x}^{n-1} f(\xi) \frac{(1+\lambda_2)^x (1+\lambda_1)^{n-\xi-1} - (1+\lambda_2)^{n+x-\xi-1}}{(\lambda_2 - \lambda_1)\left[(1+\lambda_2)^n - (1+\lambda_1)^n\right]}$$

or equivalently

$$y(x) = \frac{1}{(\lambda_2 - \lambda_1)\Delta} \left\{ (1+\lambda_2)^n (1+\lambda_1)^x \int_0^x \left[(1+\lambda_2)^{-\xi-1} - (1+\lambda_1)^{-\xi-1}\right] f(\xi) \right.$$

$$- (1+\lambda_1)^n (1+\lambda_2)^x \int_0^x \left[(1+\lambda_2)^{-\xi-1} - (1+\lambda_1)^{-\xi-1}\right] f(\xi)$$

$$+ (1+\lambda_1)^x \int_x^n \left[(1+\lambda_2)^{n-\xi-1} - (1+\lambda_1)^{n-\xi-1}\right] f(\xi)$$

$$\left. - (1+\lambda_2)^x \int_x^n \left[(1+\lambda_2)^{n-\xi-1} - (1+\lambda_1)^{n-\xi-1}\right] f(\xi) \right\} .$$

In this way, for the solution of the boundary value problem (3.45)–(3.46) we obtain the following

$$
\begin{aligned}
y(x) = {} & \frac{(1+\lambda_2)^n(1+\lambda_1)^x - (1+\lambda_2)^x(1+\lambda_1)^n}{(\lambda_2 - \lambda_1)\left[(1+\lambda_2)^n - (1+\lambda_1)^n\right]} \\
& \times \int_0^x \left[(1+\lambda_2)^{-\xi-1} - (1+\lambda_1)^{-\xi-1}\right] f(\xi) \\
& + \frac{(1+\lambda_1)^x - (1+\lambda_2)^x}{(\lambda_2 - \lambda_1)\left[(1+\lambda_2)^n - (1+\lambda_1)^n\right]} \\
& \times \int_x^n \left[(1+\lambda_2)^{n-\xi-1} - (1+\lambda_1)^{n-\xi-1}\right] f(\xi) \;.
\end{aligned}
\tag{3.48}
$$

From this expression it is immediately clear that the boundary condition holds.

Next we are going to seek the corresponding general solution for (3.45) such that the discrete integral in (3.47) is also extendable on $[x, n]$. For this purpose, we pick up the general solution of the corresponding homogeneous equation of (3.45) i.e.,

$$
y(x) = \sum_{p=1}^{2} C_{p2}(1+\lambda_p)^x \qquad C_{p2} \;\; (p = 1, 2)
$$

and seek the general solution of (3.45) as

$$
y(x) = \sum_{p=1}^{2} C_{p2}(x)(1+\lambda_p)^x \;.
\tag{3.49}
$$

Now to calculate the functions C'_{p2} $(p = 1, 2)$ we use the method of the variation of parameters to form the system of equations

$$
\begin{cases}
\displaystyle\sum_{p=1}^{2} C'_{p2}(x)(1+\lambda_p)^{x+1} & = 0 \\[2.5em]
\displaystyle\sum_{p=1}^{2} C'_{p2}(x)\lambda_p(1+\lambda_p)^{x+1} & = f(x)
\end{cases}
$$

in terms of $C'_{p2}(x)$ $(p = 1, 2)$.

Solving these system of equations we have

$$\begin{cases} C'_{12}(x) & = \frac{(1+\lambda_1)^{-x-1}}{\lambda_1-\lambda_2}\ f(x) \\[2mm] C'_{22}(x) & = \frac{(1+\lambda_2)^{-x-1}}{\lambda_2-\lambda_1}\ f(x)\ . \end{cases} \tag{3.50}$$

By integration in $[x,n]$ we deduce that

$$C_{12}(x) = C_{12}(n) + \int_x^n \frac{(1+\lambda_1)^{-\xi-1}}{\lambda_2-\lambda_1}\ f(\xi)$$

$$C_{22}(x) = C_{22}(n) - \int_x^n \frac{(1+\lambda_2)^{-\xi-1}}{\lambda_2-\lambda_1}\ f(\xi)\ .$$

Finally by denoting $C_{p2}(n) = C_{p2}$ and substituting the above expressions in (3.49) for the equation (3.45) we obtain the general solution of

$$y(x) = \sum_{p=1}^{2} C_{p2}(1+\lambda_p)^x + \int_x^n g_2(x-\xi)f(\xi) \tag{3.51}$$

where

$$g_2(x-\xi) = \frac{(1+\lambda_1)^{x-\xi-1} - (1+\lambda_2)^{x-\xi-1}}{\lambda_2-\lambda_1} \qquad x \le \xi < n\ . \tag{3.52}$$

By averaging the two general solutions (3.47) and (3.51) we have

$$y(x) = \sum_{p=1}^{2} C_p(1+\lambda_p)^x + \int_0^n g(x-\xi)f(\xi) \tag{3.53}$$

where $C_p = \frac{1}{2}(C_{p1}+C_{p2})\ (p=1,2)$ are arbitrary constants and the function

$$g(x-\xi) = \begin{cases} \frac{1}{2(\lambda_2-\lambda_1)}\left[(1+\lambda_2)^{x-\xi-1} - (1+\lambda_1)^{x-\xi-1}\right] & 0 \le \xi < x \\[2mm] \frac{-1}{2(\lambda_2-\lambda_1)}\left[(1+\lambda_2)^{x-\xi-1} - (1+\lambda_1)^{x-\xi-1}\right] & x \le \xi < n \end{cases} \tag{3.54}$$

is the fundamental solution of (3.45). For future purpose, we use the more appropriate expression

$$g(x-\xi) = \frac{(1+\lambda_2)^{x-\xi-1} - (1+\lambda_1)^{x-\xi-1}}{\lambda_2-\lambda_1}\ e(x-\xi) \tag{3.54_1}$$

where

$$e(x - \xi) = \begin{cases} \frac{1}{2} & x > \xi \\ \frac{-1}{2} & x \le \xi \end{cases} . \tag{3.55}$$

It is easily seen that

$$e'(x - \xi) = \begin{cases} 0 & x > \xi \\ 1 & x = \xi \\ 0 & x < \xi \end{cases} . \tag{3.55_1}$$

So according to the expressions

$$g'(x - \xi) = \frac{\lambda_2(1 + \lambda_2)^{x-\xi-1} - \lambda_1(1 + \lambda_2)^{x-\xi-1}}{\lambda_2 - \lambda_1} \, e(x - \xi)$$

and

$$g''(x - \xi) = e'(x - \xi) + \frac{\lambda_2^2(1 + \lambda_2)^{x-\xi-1} - \lambda_1^2(1 + \lambda_1)^{x-\xi-1}}{\lambda_2 - \lambda_1} \, e(x - \xi)$$

the function $g(x - \xi)$ is the fundamental solution of the equation

$$g''(x - \xi) - 2ag'(x - \xi) - bg(x - \xi) = \delta(x - \xi) \tag{3.56}$$

where $\delta(x - \xi) = e'(x - \xi)$.

Finally by imposing the boundary conditions (3.46) in (3.53) for the constants C_p $(p = 1, 2)$ we obtain the system of equations

$$\begin{cases} \sum\limits_{p=1}^{2} C_p & = -\displaystyle\int_0^n g(-\xi)f(\xi) \\ \sum\limits_{p=1}^{2} C_p(1 + \lambda_p)^n & = -\displaystyle\int_0^n g(n - \xi)f(\xi) \end{cases} .$$

By substituting the solution of this system in (3.53) for the boundary value problem (3.45)–(3.46) we obtain the solution of

$$y(x) = \int_0^n G(x, x - \xi)f(\xi) \tag{3.57}$$

where

$$G(x, x - \xi) = \frac{1}{\Delta} \begin{vmatrix} g(x - \xi) & (1 + \lambda_1)^x & (1 + \lambda_2)^x \\ g(-\xi) & 1 & 1 \\ g(n - \xi) & (1 + \lambda_1)^n & (1 + \lambda_2)^n \end{vmatrix} \qquad (3.58)$$

is called the kernel of the solution.

4   Boundary Value Problems

4

Abstract. The theory of discrete value problems and their connections to the spectral theory of operators are discussed in this chapter. These problems can be interpreted as the discrete analog of the mathematical physics equations in one dimension. These include auxiliary problem scheme, coincidence condition of a boundary value problem on its auxiliary problem, the type of problems coinciding with their auxiliary problems, the necessary condition for existence of the boundary value problem and finally the boundary value problem on an infinite domain.

4.1   Auxiliary problem scheme

Consider the boundary value problem

$$Ly = ay''(x) = by'(x) + cy(x) = f(x) \qquad 0 \le x \le n \tag{4.1}$$

$$L_j y = \alpha_{j1} y'(0) + \beta_{j1} y'(n) + \alpha_j y(0) + \beta_j y(n) = 0 \qquad (j = 1, 2) \tag{4.2}$$

where $a, b, c, \alpha_{j1}, \beta_{j1}, \alpha_j, \beta_j$ $(j = 1, 2)$ are given constants and $f(x)$ is an arbitrary function. For the auxiliary problem scheme, we multiply the right-hand side of (4.1) by $z''(x) + 2z'(x) + z(x)$ and integrate the resulting expression

---

on $[0, n)$. In fact we have

$$\int_0^n Ly(z'' + 2z' + z) = \int_0^n (ay'' + by' + cy)(z'' + 2z' + z)$$

$$= \int_0^n \Big\{ a\left[(y'z')' - y''z' - y'z''\right]$$

$$+ 2a\left[(y'z)' - y''z - y'z'\right] + ay''z + b\left[(yz')' - y'z' - yz''\right]$$

$$+ 2b\left[(yz)' - y'z - yz'\right] + by'z + cyz'' + 2cyz' + cyz\Big\}$$

$$= \int_0^n \Big\{ \left[ay'z' + ay'z - ayz' + byz' + byz\right]'$$

$$+ (a - b + c)yz'' - (b - 2c)yz' + cyz\Big\}$$

$$= \int_0^n y\left[(a - b + c)z'' - (b - 2c)z' + cz\right]$$

$$+ \left[ay'(z' + z) - (a - b)yz' + byz\right]_0^n \ .$$

$$(4.3)$$

In this way, corresponding to Ly in (4.1) we obtain the operator L^*y as

$$L^*y \equiv (a - b + c)z''(x) - (b - 2c)z'(x) + cz(x) = g(x) \qquad 0 \le x \le n \ . \ (4.4)$$

It is appropriate now to find the boundary condition of the auxiliary problem. For this purpose we examine the system of equations of

$$\alpha_{j1}y'(0) + \beta_{j1}y'(n) + \alpha_j y(0) + \beta_j y(n) = L_j y \qquad (j = 1, 2, 3, 4) \qquad (4.5)$$

where the equations under consideration for $j = 1, 2$ are the boundary conditions for the expressions (4.2) and for $j = 3, 4$ are arbitrary expressions. Suppose that

$$\Delta = \begin{vmatrix} \alpha_{11} & \beta_{11} & \alpha_1 & \beta_1 \\ \alpha_{21} & \beta_{21} & \alpha_2 & \beta_2 \\ \alpha_{31} & \beta_{31} & \alpha_3 & \beta_3 \\ \alpha_{41} & \beta_{41} & \alpha_4 & \beta_4 \end{vmatrix} \ne 0 \ , \qquad (4.6)$$

so by Cramer's rule for the system (4.5) we obtain

$$
\begin{cases}
y'(0) &= \frac{1}{\Delta}\left(L_1 y \cdot A_{11} + L_2 y \cdot A_{21} + L_3 y \cdot A_{31} + L_4 y \cdot A_{41}\right) \\
y'(n) &= \frac{1}{\Delta}\left(L_1 y \cdot B_{11} + L_2 y \cdot B_{21} + L_3 y \cdot B_{31} + L_4 y \cdot B_{41}\right) \\
y(0) &= \frac{1}{\Delta}\left(L_1 y \cdot A_1 + L_2 y \cdot A_2 + L_3 y \cdot A_3 + L_4 y \cdot A_4\right) \\
y(n) &= \frac{1}{\Delta}\left(L_1 y \cdot B_1 + L_2 y \cdot B_2 + L_3 y \cdot B_3 + L_4 y \cdot B_4\right)
\end{cases}
\tag{4.7}
$$

where A_j, B_{j1}, A_{j1}, B_j $(j=1,\ldots,4)$ are corresponding cofactors of $\alpha_j, \beta_{j1}, \alpha_{j1}, \beta_j$ $(j=1,\ldots,4)$ in the above determinant respectively. Now according to the convention made in (4.4) by using the expression (4.7) we write down the discrete analog of Lagrange formula in (4.3), in fact

$$
\int_0^n Ly(z'' + 2z' + z) = \int_0^n yL^* z
$$
$$
+ a\left(z'(n) + z(n)\right)\frac{1}{\Delta}\left(L_1 y B_{11} + L_2 y B_{21} + L_3 y B_{31} + L_4 y B_{41}\right)
$$
$$
+ [(b-a)z'(n) + bz(n)]\frac{1}{\Delta}\left(L_1 y B_1 + L_2 y B_2 + L_3 y B_3 + L_4 y B_4\right)
$$
$$
- a\left(z'(0) + z(0)\right)\frac{1}{\Delta}\left(L_1 y A_{11} + L_2 y A_{21} + L_3 y A_{31} + L_4 y A_{41}\right)
$$
$$
- [(b-a)z'(0) + bz(0)]\frac{1}{\Delta}\left(L_1 y A_1 + L_2 y A_2 + L_3 y A_3 + L_4 y A_4\right).
$$

If we accept the equality $L_1 y = L_2 y = 0$, i.e., the function y satisfies the condition (4.2) then it can be easily seen that in order the expression outside of the integral notation vanishes independent of $L_3 y$ and $L_4 y$, the following boundary condition must hold for the function z

$$
a\left(z'(n) + z(n)\right) B_{j1} + [(b-a)z'(n) + Bz(n)] B_j
$$
$$
- a\left(z'(0) + z(0)\right) A_{j1} - [(b-a)z'(0) + bz(0)] A_j = 0 \quad (j=3,4) \tag{4.8}
$$

Therefore the corresponding auxiliary problem of (4.1)–(4.2) is in the form of (4.2)–(4.8).

4.2   Coincidence condition of a boundary value problem on its auxiliary problem

In this section we are going to examine the conditions on which the boundary value problem of (4.1)–(4.2) coincides with its auxiliary problem. To this end, we compare these two problems in details. First let us determine the coincidence condition of (4.1) on (4.4). By comparing these two equations we realize that in order to have a coincidence condition, the following must be true

$$\begin{cases} a - b + c & = a \\ -(b - 2c) & = b \\ c & = c \ . \end{cases}$$

Clearly these conditions are equivalent to the single condition

$$b = c \ . \tag{4.9}$$

We say that under this last condition, the equation (4.1) coincide with its auxiliary problem (4.4).

Next we compare the boundary conditions (4.2) and (4.8), for this purpose we have

$$\begin{cases} -aA_{2+j1} - (b - a)A_{2+j} & = \alpha_{j1} \\ aB_{2+j1} + (b - a)B_{2+j} & = \beta_{j1} \\ -aA_{2+j1} - bA_{2+j} & = \alpha_j \\ aB_{2+j1} + bB_{2+j} & = \beta_j \qquad (j = 1, 2) \ . \end{cases}$$

Solving this system we get

$$\begin{cases} A_{2+j} & = \frac{\alpha_{j1} - \alpha_j}{a} \\ B_{2+j} & = \frac{\beta_j - \beta_{j1}}{a} \\ A_{2+j1} & = \frac{a\alpha_j + b(\alpha_{j1} - \alpha_j)}{a^2} \\ B_{2+j1} & = \frac{a\beta_j - b(\beta_j - \beta_{j1})}{a^2} \qquad (j = 1, 2) \end{cases} \tag{4.10}$$

Now we calculate the corresponding cofactors of (4.10) in (4.6), that is

$$A_3 = \begin{vmatrix} \alpha_{11} & \beta_{11} & \beta_1 \\ \alpha_{21} & \beta_{21} & \beta_2 \\ \alpha_{41} & \beta_{41} & \beta_4 \end{vmatrix} \qquad A_4 = \begin{vmatrix} \alpha_{11} & \beta_{11} & \beta_1 \\ \alpha_{21} & \beta_{21} & \beta_2 \\ \alpha_{31} & \beta_{31} & \beta_3 \end{vmatrix}$$

$$B_3 = \begin{vmatrix} \alpha_{11} & \beta_{11} & \alpha_1 \\ \alpha_{21} & \beta_{21} & \alpha_2 \\ \alpha_{41} & \beta_{41} & \alpha_4 \end{vmatrix} \qquad B_4 = \begin{vmatrix} \alpha_{11} & \beta_{11} & \alpha_1 \\ \alpha_{21} & \beta_{21} & \alpha_2 \\ \alpha_{31} & \beta_{31} & \alpha_3 \end{vmatrix}$$

$$A_{31} = \begin{vmatrix} \beta_{11} & \alpha_1 & \beta_1 \\ \beta_{21} & \alpha_2 & \beta_2 \\ \beta_{41} & \alpha_4 & \beta_4 \end{vmatrix} \qquad A_{41} = -\begin{vmatrix} \beta_{11} & \alpha_1 & \beta_1 \\ \beta_{21} & \alpha_2 & \beta_2 \\ \beta_{31} & \alpha_3 & \beta_3 \end{vmatrix}$$

$$B_{31} = -\begin{vmatrix} \alpha_{11} & \alpha_1 & \beta_1 \\ \alpha_{21} & \alpha_2 & \beta_2 \\ \alpha_{41} & \alpha_4 & \beta_4 \end{vmatrix} \qquad B_{41} = \begin{vmatrix} \alpha_{11} & \alpha_1 & \beta_1 \\ \alpha_{21} & \alpha_2 & \beta_2 \\ \alpha_{31} & \alpha_3 & \beta_3 \end{vmatrix}$$

To determine the constants of the two last expressions in (4.5) we substitute the above values in (4.10) and form the related linear system of equations. Because of non-homogeneousness of this system it is necessary to verify the existence of the solution. By a little attention, we observe that this system of equations reduces into two independent system. We discuss one of these systems

$$\begin{cases} \alpha_{41}\begin{vmatrix} \beta_{11} & \beta_1 \\ \beta_{21} & \beta_2 \end{vmatrix} - \beta_{41}\begin{vmatrix} \alpha_{11} & \beta_1 \\ \alpha_{21} & \beta_2 \end{vmatrix} + \beta_4\begin{vmatrix} \alpha_{11} & \beta_{11} \\ \alpha_{21} & \beta_{21} \end{vmatrix} & = \frac{\alpha_{11}-\alpha_1}{a} \\[4mm] \alpha_{41}\begin{vmatrix} \beta_{11} & \alpha_1 \\ \beta_{21} & \alpha_2 \end{vmatrix} - \beta_{41}\begin{vmatrix} \alpha_{11} & \alpha_1 \\ \alpha_{21} & \alpha_2 \end{vmatrix} + \alpha_4\begin{vmatrix} \alpha_{11} & \beta_{11} \\ \alpha_{21} & \beta_{21} \end{vmatrix} & = \frac{\beta_{11}-\beta_1}{a} \\[4mm] \beta_{41}\begin{vmatrix} \alpha_1 & \beta_1 \\ \alpha_2 & \beta_2 \end{vmatrix} - \alpha_4\begin{vmatrix} \beta_{11} & \beta_1 \\ \beta_{21} & \beta_2 \end{vmatrix} + \beta_4\begin{vmatrix} \beta_{11} & \alpha_1 \\ \beta_{21} & \alpha_2 \end{vmatrix} & = -\frac{a\alpha_1+b(\alpha_{11}-\alpha_1)}{a^2} \\[4mm] \alpha_{41}\begin{vmatrix} \alpha_1 & \beta_1 \\ \alpha_2 & \beta_2 \end{vmatrix} - \alpha_4\begin{vmatrix} \alpha_{11} & \beta_1 \\ \alpha_{21} & \beta_2 \end{vmatrix} + \beta_4\begin{vmatrix} \alpha_{11} & \alpha_1 \\ \alpha_{21} & \alpha_2 \end{vmatrix} & = -\frac{a\beta_1-b(\beta_1-\beta_{11})}{a^2} \, . \end{cases} \tag{4.11}$$

If the condition

$$\Delta_0 = \begin{vmatrix} \alpha_{11} & \beta_{11} \\ \alpha_{21} & \beta_{21} \end{vmatrix} \neq 0 \tag{4.12}$$

4 Boundary Value Problems

Discrete Calculus By Analogy   87

holds, then from the two homogeneous equations of (4.11) we obtain the following

$$
\begin{cases}
\alpha_4 = \dfrac{\beta_{11}-\beta_1}{a\Delta_0} - \dfrac{1}{\Delta_0}\begin{vmatrix}\beta_{11} & \alpha_1 \\ \beta_{21} & \alpha_2\end{vmatrix}\alpha_{41} + \dfrac{1}{\Delta_0}\begin{vmatrix}\alpha_{11} & \alpha_1 \\ \alpha_{21} & \alpha_2\end{vmatrix}\beta_{41} \\[4mm]
\beta_4 = \dfrac{\alpha_{11}-\alpha_1}{a\Delta_0} - \dfrac{1}{\Delta_0}\begin{vmatrix}\beta_{11} & \beta_1 \\ \beta_{21} & \beta_2\end{vmatrix}\alpha_{41} + \dfrac{1}{\Delta_0}\begin{vmatrix}\alpha_{11} & \beta_1 \\ \alpha_{21} & \beta_2\end{vmatrix}\beta_{41} \; .
\end{cases}
\tag{4.13}
$$

By substituting these expressions in the two remaining equations of (4.11) we get

$$
\begin{vmatrix}\alpha_1 & \beta_1 \\ \alpha_2 & \beta_2\end{vmatrix}\beta_{41} - \begin{vmatrix}\beta_{11} & \beta_1 \\ \beta_{21} & \beta_2\end{vmatrix}\times
$$

$$
\times\left(\dfrac{\beta_{11}-\beta_1}{a\Delta_0} - \dfrac{1}{\Delta_0}\begin{vmatrix}\beta_{11} & \alpha_1 \\ \beta_{21} & \alpha_2\end{vmatrix}\alpha_{41} + \dfrac{1}{\Delta_0}\begin{vmatrix}\alpha_{11} & \alpha_1 \\ \alpha_{21} & \alpha_2\end{vmatrix}\beta_{41}\right) +
$$

$$
+\begin{vmatrix}\beta_{11} & \alpha_1 \\ \beta_{21} & \alpha_2\end{vmatrix}\left(\dfrac{\alpha_{11}-\alpha_1}{a\Delta_0} - \dfrac{1}{\Delta_0}\begin{vmatrix}\beta_{11} & \beta_1 \\ \beta_{21} & \beta_2\end{vmatrix}\alpha_{41} + \dfrac{1}{\Delta_0}\begin{vmatrix}\alpha_{11} & \beta_1 \\ \alpha_{21} & \beta_2\end{vmatrix}\beta_{41}\right)
$$

$$
= -\dfrac{a\alpha_1 + b(\alpha_{11}-\alpha_1)}{a^2}
$$

$$
\begin{vmatrix}\alpha_1 & \beta_1 \\ \alpha_2 & \beta_2\end{vmatrix}\alpha_{41} - \begin{vmatrix}\alpha_{11} & \beta_1 \\ \alpha_{21} & \beta_2\end{vmatrix}\times
$$

$$
\times\left(\dfrac{\beta_{11}-\beta_1}{a\Delta_0} - \dfrac{1}{\Delta_0}\begin{vmatrix}\beta_{11} & \alpha_1 \\ \beta_{21} & \alpha_2\end{vmatrix}\alpha_{41} + \dfrac{1}{\Delta_0}\begin{vmatrix}\alpha_{11} & \alpha_1 \\ \alpha_{21} & \alpha_2\end{vmatrix}\beta_{41}\right) +
$$

$$
+\begin{vmatrix}\alpha_{11} & \alpha_1 \\ \alpha_{21} & \alpha_2\end{vmatrix}\left(\dfrac{\alpha_{11}-\alpha_1}{a\Delta_0} - \dfrac{1}{\Delta_0}\begin{vmatrix}\beta_{11} & \beta_1 \\ \beta_{21} & \beta_2\end{vmatrix}\alpha_{41} + \dfrac{1}{\Delta_0}\begin{vmatrix}\alpha_{11} & \beta_1 \\ \alpha_{21} & \beta_2\end{vmatrix}\beta_{41}\right)
$$

$$
= -\dfrac{a\beta_1 - b(\beta_1-\beta_{11})}{a^2} \; .
$$

Now consider the following conditions

$$
\begin{cases}
-a(\beta_{11} - \beta_1)(\beta_{11}\beta_2 - \beta_{21}\beta_1) + a(\alpha_{11} - \alpha_1)(\beta_{11}\alpha_2 - \beta_{21}\alpha_1) = \\
-(\alpha_{11}\beta_{21} - \alpha_{21}\beta_{11})\left[a\alpha_1 + b(\alpha_{11} - \alpha_1)\right] \\
-a(\beta_{11} - \beta_1)(\alpha_{11}\beta_2 - \alpha_{21}\beta_1) + a(\alpha_{11} - \alpha_1)(\alpha_{11}\alpha_2 - \alpha_{21}\alpha_1) = \\
-(\alpha_{11}\beta_{21} - \alpha_{21}\beta_{11})\left[a\beta_1 - b(\beta_1 - \beta_{11})\right] \ .
\end{cases}
\tag{4.14}
$$

Similarly the second system of (4.10) reduces to

$$
\begin{cases}
\alpha_{31}\begin{vmatrix} \beta_{11} & \beta_1 \\ \beta_{21} & \beta_2 \end{vmatrix} - \beta_{31}\begin{vmatrix} \alpha_{11} & \beta_1 \\ \alpha_{21} & \beta_2 \end{vmatrix} + \beta_3\begin{vmatrix} \alpha_{11} & \beta_{11} \\ \alpha_{21} & \beta_{21} \end{vmatrix} &= -\frac{\alpha_{21} - \alpha_2}{a} \\[2ex]
\alpha_{31}\begin{vmatrix} \beta_{11} & \alpha_1 \\ \beta_{21} & \alpha_2 \end{vmatrix} - \beta_{31}\begin{vmatrix} \alpha_{11} & \alpha_1 \\ \alpha_{21} & \alpha_2 \end{vmatrix} + \alpha_3\begin{vmatrix} \alpha_{11} & \beta_{11} \\ \alpha_{21} & \beta_{21} \end{vmatrix} &= \frac{\beta_2 - \beta_{21}}{a} \\[2ex]
\beta_{31}\begin{vmatrix} \alpha_1 & \beta_1 \\ \alpha_2 & \beta_2 \end{vmatrix} - \alpha_3\begin{vmatrix} \beta_{11} & \beta_1 \\ \beta_{21} & \beta_2 \end{vmatrix} + \beta_3\begin{vmatrix} \beta_{11} & \alpha_1 \\ \beta_{21} & \alpha_2 \end{vmatrix} &= \frac{a\alpha_2 + b(\alpha_{21} - \alpha_2)}{a^2} \\[2ex]
\alpha_{31}\begin{vmatrix} \alpha_1 & \beta_1 \\ \alpha_2 & \beta_2 \end{vmatrix} - \alpha_3\begin{vmatrix} \alpha_{11} & \beta_1 \\ \alpha_{21} & \beta_2 \end{vmatrix} + \beta_3\begin{vmatrix} \alpha_{11} & \alpha_1 \\ \alpha_{21} & \alpha_2 \end{vmatrix} &= \frac{a\beta_2 - b(\beta_2 - \beta_{21})}{a^2} \ .
\end{cases}
\tag{4.15}
$$

Now the values of α_3 and β_3 are obtained via the first two equations of (4.12), that is

$$
\beta_3 = -\frac{\alpha_{21} - \alpha_2}{a\Delta_0} - \frac{1}{\Delta_0}\begin{vmatrix} \beta_{11} & \beta_1 \\ \beta_{21} & \beta_2 \end{vmatrix}\alpha_{31} + \frac{1}{\Delta_0}\begin{vmatrix} \alpha_{11} & \beta_1 \\ \alpha_{21} & \beta_2 \end{vmatrix}\beta_{31}
$$

$$
\alpha_3 = \frac{\beta_2 - \beta_{21}}{a\Delta_0} - \frac{1}{\Delta_0}\begin{vmatrix} \beta_{11} & \alpha_1 \\ \beta_{21} & \alpha_2 \end{vmatrix}\alpha_{31} + \frac{1}{\Delta_0}\begin{vmatrix} \alpha_{11} & \alpha_1 \\ \alpha_{21} & \alpha_2 \end{vmatrix}\beta_{31} \ .
$$

By substituting the values in the last two equations we have

$$
\begin{vmatrix} \alpha_1 & \beta_1 \\ \alpha_2 & \beta_2 \end{vmatrix} \beta_{31} - \begin{vmatrix} \beta_{11} & \beta_1 \\ \beta_{21} & \beta_2 \end{vmatrix} \times
$$

$$
\times \left(\frac{\beta_2 - \beta_{21}}{a\Delta_0} - \frac{1}{\Delta_0} \begin{vmatrix} \beta_{11} & \alpha_1 \\ \beta_{21} & \alpha_2 \end{vmatrix} \alpha_{31} + \frac{1}{\Delta_0} \begin{vmatrix} \alpha_{11} & \alpha_1 \\ \alpha_{21} & \alpha_2 \end{vmatrix} \beta_{31} \right) +
$$

$$
+ \begin{vmatrix} \beta_{11} & \alpha_1 \\ \beta_{21} & \alpha_2 \end{vmatrix} \left(\frac{\alpha_2 - \alpha_{21}}{a\Delta_0} - \frac{1}{\Delta_0} \begin{vmatrix} \beta_{11} & \beta_1 \\ \beta_{21} & \beta_2 \end{vmatrix} \alpha_{31} + \frac{1}{\Delta_0} \begin{vmatrix} \alpha_{11} & \beta_1 \\ \alpha_{21} & \beta_2 \end{vmatrix} \beta_{31} \right) =
$$

$$
= \frac{a\alpha_2 + b(\alpha_{21} - \alpha_2)}{a^2}
$$

$$
\begin{vmatrix} \alpha_1 & \beta_1 \\ \alpha_2 & \beta_2 \end{vmatrix} \alpha_{31} - \begin{vmatrix} \alpha_{11} & \beta_1 \\ \alpha_{21} & \beta_2 \end{vmatrix} \times
$$

$$
\times \left(\frac{\beta_2 - \beta_{21}}{a\Delta_0} - \frac{1}{\Delta_0} \begin{vmatrix} \beta_{11} & \alpha_1 \\ \beta_{21} & \alpha_2 \end{vmatrix} \alpha_{31} + \frac{1}{\Delta_0} \begin{vmatrix} \alpha_{11} & \alpha_1 \\ \alpha_{21} & \alpha_2 \end{vmatrix} \beta_{31} \right) +
$$

$$
+ \begin{vmatrix} \alpha_{11} & \alpha_1 \\ \alpha_{21} & \alpha_2 \end{vmatrix} \left(\frac{\alpha_2 - \alpha_{21}}{a\Delta_0} - \frac{1}{\Delta_0} \begin{vmatrix} \beta_{11} & \beta_1 \\ \beta_{21} & \beta_2 \end{vmatrix} \alpha_{31} + \frac{1}{\Delta_0} \begin{vmatrix} \alpha_{11} & \beta_1 \\ \alpha_{21} & \beta_2 \end{vmatrix} \beta_{31} \right) =
$$

$$
= \frac{a\beta_2 - b(\beta_2 - \beta_{21})}{a^2} \; .
$$

These expressions imply the following equalities

$$
\begin{cases}
-a(\beta_2 - \beta_{21})(\beta_{11}\beta_2 - \beta_{21}\beta_1) + a(\alpha_2 - \alpha_{21})(\beta_{11}\alpha_2 - \beta_{21}\alpha_1) \\
= (\alpha_{11}\beta_{21} - \alpha_{21}\beta_{11}) [a\alpha_2 + b(\alpha_{21} - \alpha_2)] \\
-a(\beta_2 - \beta_{21})(\alpha_{11}\beta_2 - \alpha_{21}\beta_1) + a(\alpha_2 - \alpha_{21})(\alpha_{11}\alpha_2 - \alpha_{21}\alpha_1) \\
= (\alpha_{11}\beta_{21} - \alpha_{21}\beta_{11}) [a\beta_2 - b(\beta_2 - \beta_{21})] \; .
\end{cases}
\tag{4.16}
$$

In this way, for coincidence of boundary conditions (4.2) and (4.8), the equalities (4.14) and (4.16) must hold.

Remark.   It is easily seen that the equalities (4.14) and (4.16) are in turn obtained from the sum of the products of entries of the first two rows to the corresponding cofactors of the last two rows of the 4×4 determinant appeared in (4.10). Therefore, these expressions is in the form

$$
\begin{cases}
\alpha_{11}A_{31} + \beta_{11}B_{31} + \alpha_1 A_3 + \beta_1 B_3 & = 0 \\
\alpha_{21}A_{31} + \beta_{21}B_{31} + \alpha_2 A_3 + \beta_2 B_3 & = 0 \\
\alpha_{11}A_{41} + \beta_{11}B_{41} + \alpha_1 A_4 + \beta_1 B_4 & = 0 \\
\alpha_{21}A_{41} + \beta_{21}B_{41} + \alpha_2 A_4 + \beta_2 B_4 & = 0 \, .
\end{cases}
$$

Now according to (4.10) these expressions reduce to

$$
-\alpha_{11}\frac{a\alpha_1 + b(\alpha_{11} - \alpha_1)}{a^2} + \beta_{11}\frac{a\beta_1 - b(\beta_1 - \beta_{11})}{a^2} + \alpha_1\frac{\alpha_{11} - \alpha_1}{a} + \beta_1\frac{\beta_1 - \beta_{11}}{a} = 0
$$

$$
-\alpha_{21}\frac{a\alpha_1 + b(\alpha_{11} - \alpha_1)}{a^2} + \beta_{21}\frac{a\beta_1 - b(\beta_1 - \beta_{11})}{a^2} + \alpha_2\frac{\alpha_{11} - \alpha_1}{a} + \beta_2\frac{\beta_1 - \beta_{11}}{a} = 0
$$

$$
-\alpha_{11}\frac{a\alpha_2 + b(\alpha_{21} - \alpha_2)}{a^2} + \beta_{11}\frac{a\beta_2 - b(\beta_2 - \beta_{21})}{a^2} + \alpha_1\frac{\alpha_{21} - \alpha_2}{a} + \beta_1\frac{\beta_2 - \beta_{21}}{a} = 0
$$

$$
-\alpha_{21}\frac{a\alpha_2 + b(\alpha_{21} - \alpha_2)}{a^2} + \beta_{21}\frac{a\beta_2 - b(\beta_2 - \beta_{21})}{a^2} + \alpha_2\frac{\alpha_{21} - \alpha_2}{a} + \beta_2\frac{\beta_2 - \beta_{21}}{a} = 0 \, .
$$

$$(4.17)$$

By this way, the coincidence condition of the problem (4.1) and (4.2) with its auxiliary problem will be in the form of (4.9) and (4.17).

4.3   Problems which coincide with their auxiliary problems

From the idea used in previous section we observe that the boundary value problem

$$ay''(x) + by'(x) + by(x) = f(x) \qquad 0 \le x < n$$

$$\alpha_{j1}y'(0) + \beta_{j1}y'(n) + \alpha_j y(0) + \beta_j y(n) = 0 \qquad (j = 1, 2)$$

under the conditions

$$\alpha_{21}\alpha_1 - \beta_{21}\beta_1 = \alpha_{11}\alpha_2 - \beta_{11}\beta_2$$

$$(\alpha_{11}\alpha_2 - \beta_{11}\beta_2 + \beta_{11}\beta_{21} - \alpha_{11}\alpha_{21})b = (\alpha_1\alpha_2 - \beta_1\beta_2)a$$

$$(\alpha_{11}\alpha_1 - \beta_{11}\beta_1 + \beta_{11}^2 - \alpha_{11}^2)b = (\alpha_1^2 - \beta_1^2)a$$

$$(\alpha_{21}\alpha_2 - \beta_{21}\beta_2 + \beta_{21}^2 - \alpha_{21}^2)b = (\alpha_2^2 - \beta_2^2)a$$

(4.18)

coincides with its auxiliary problem. In particular case, by assuming

$$a = \alpha_1 = \beta_1 = 1 \qquad \alpha_{11} = \beta_{11} = 0$$

$$\alpha_{21} = \beta_{21} = 1 \qquad \alpha_2 = \beta_2 = 0$$

we see that the condition (4.18) holds, so we have the following problem

$$y''(x) + by'(x) + by(x) = f(x) \qquad 0 \le x < n$$

$$y(0) + y(n) = 0$$

$$y'(0) + y'(n) = 0 \ .$$

It can be easily shown that the following problem coincides with its auxiliary problem

$$y''(x) + by'(x) + by(x) = f(x) \qquad 0 \le x < n$$

$$y(0) = y(n)$$

$$y'(0) = y'(n) \ .$$

From the conditions (4.18) we can determine the four coefficients appeared in boundary conditions. So the number of these problems can be increased, in fact by assuming

$$a = -2, \ b = 1, \ \alpha_1 = -1, \ \alpha_{11} = 1, \ \beta_{11} = \beta_1 = \alpha_{21} = \alpha_2 = 0, \ \beta_2 = -1, \ \beta_{21} = 1$$

we obtain the following problem

$$-2y''(x) + y'(x) + y(x) = f(x) \qquad 0 \le x < n$$

$$y'(0) = y(0)$$

$$y'(n) = y(n) \ .$$

In order to coincidence with its auxiliary problem, this problem must have a separate boundary conditions, i.e., there must be one boundary condition at 0 and another one at n. By dividing both sides of the equation by a we will have only one equation with coefficient b/a, if we can find the four coefficients appearing in (4.18), substitute them in the boundary conditions and divide the expression by one of the non-zero coefficients, then the resulting problem will depend only on three parameters. Consequently the set of all problems of second order which coincide with their auxiliary problems, explicitly depend on three independent parameters.

4.4   The necessary condition for existence of the boundary value problem

Let us suppose that in the boundary value problem (4.1) and (4.2) we have $a = 1$ and $b = c$. Also instead of the fundamental solution $z(x)$ in (4.3) we write $z(x - \xi)$ (in this case (4.1) coincides with (4.2)) then we have

$$\int_0^n f(x) \left[z_x''(x - \xi) + 2z_x'(x - \xi) + z(x - \xi) \right] dx$$
$$+ y'(0) \left[z_x'(-\xi) + z(-\xi) \right] - (1 - b)y(0)z_x'(-\xi) + by(0)z(-\xi)$$
$$- y'(n) \left[z_x'(n - \xi) + z(n - \xi) \right] + (1 - b)y(n)z_x'(n - \xi) - by(n)z(n - \xi)$$
$$= \begin{cases} y(\xi) & 0 \le \xi \le n - 1 \\ 0 & \xi \ge n \end{cases}$$

$$(4.19)$$

where $z(x - \xi)$ is the solution of

$$z_x''(x - \xi) + bz_x'(x - \xi) + bz(x - \xi) = \delta(x - \xi)$$

and δ is the same function in (3.55_1).

By taking $\xi = 0$, $\xi = n$ in (4.19) we obtain the following

$$y'(0)\left[z'_x(0) + z(0)\right] - (1-b)y(0)z'_x(0) + by(0)z(0) - y'(n)\times$$
$$\times \left[z'_x(n) + z(n)\right] + (1-b)y(n)z'_x(n) - by(n)z(n)+$$
$$+ \int_0^n f(x)\left[z''_x(x) + 2z'_x(x) + z(x)\right] = y(0)$$
$$y'(0)\left[z'_x(-n) + z(-n)\right] - (1-b)y(0)z'_x(-n) + by(0)z(-n)-$$
$$- y'(n)\left[z'_x(0) + z(0)\right] + (1-b)y(n)z'_x(0) - by(n)z(0)+$$
$$+ \int_0^n f(x)\left[z''_x(x-n) + 2z'_x(x-n) + z(x-n)\right] = 0 \ .$$

$$(4.20)$$

These expressions are the above necessary conditions, to put it other way, if at least one of the conditions in (4.2) contradicts either one of the conditions in (4.20) or their linear combinations, then the problem (4.1) and (4.2) has no solution. In this stage, we examine the necessary conditions of (4.20) in the following example.

In problem (3.45) and (3.46) we assume that $a = 1$, $b = 2$, that is we think of the equation

$$y''(x) - 2y'(x) - 2y(x) = f(x) \qquad 0 \le x \le n$$

as a boundary value problem, (The boundary conditions will be chosen later) then as we have seen in (3.54_1), the fundamental solution is the following:

$$z(x - \xi) = \frac{(\lambda_2 + 1)^{x-\xi-1} - (\lambda_1 + 1))^{x-\xi-1}}{\lambda_2 - \lambda_1}e(x - \xi)$$

where

$$\lambda_j = 1 + (-1)^j\sqrt{3} \qquad (j = 1, 2) \ .$$

Finally by considering the equality

$$z''_x(x - \xi) = \delta(x - \xi) + 2z'_x(x - \xi) + 2z(x - \xi)$$

the first expression in (4.20) $(b = -2)$ becomes

$$
y'(0)\left[\frac{\lambda_2(\lambda_2 + 1)^{-1} - \lambda_1(\lambda_1 + 1)^{-1}}{\lambda_2 - \lambda_1} \, e(0) +\right.
$$

$$
\left. + \frac{(\lambda_2 + 1)^{-1} - (\lambda_1 + 1)^{-1}}{\lambda_2 - \lambda_1} \, e(0)\right] -
$$

$$
- 3y(0)\frac{\lambda_2(\lambda_2 + 1)^{-1} - \lambda_1(\lambda_1 + 1)^{-1}}{\lambda_2 - \lambda_1} \, e(0) -
$$

$$
- 2y(0)\frac{(\lambda_2 + 1)^{-1} - (\lambda_1 + 1)^{-1}}{\lambda_2 - \lambda_1} \, e(0) -
$$

$$
- y'(n)\left[\frac{\lambda_2(\lambda_2 + 1)^{n-1} - \lambda_1(\lambda_1 + 1)^{n-1}}{\lambda_2 - \lambda_1} \, e(n) +\right. \tag{4.21}
$$

$$
\left. + \frac{(\lambda_2 + 1)^{n-1} - (\lambda_1 + 1)^{n-1}}{\lambda_2 - \lambda_1} \, e(n)\right] +
$$

$$
+ 3y(n)\frac{\lambda_1(\lambda_2 + 1)^{n-1} - \lambda_1(\lambda_1 + 1)^{n-1}}{\lambda_2 - \lambda_1} \, e(n) +
$$

$$
+ 2y(n)\frac{(\lambda_2 + 1)^{n-1} - (\lambda_1 + 1)^{n-1}}{\lambda_2 - \lambda_1} \, e(n) +
$$

$$
+ \int_0^n f(x)\left[\delta(x - \xi) + 4z_x'(x - \xi) + 3z(x - \xi)\right] = y(0) \, .
$$

By assuming $f \equiv 0$ in this case, we have

$$
-\frac{1}{2}y'(0)\frac{\lambda_2(\lambda_1 + 1) - \lambda_1(\lambda_2 + 1)}{(\lambda_2 - \lambda_1)(\lambda_2 + 1)(\lambda_1 + 1)} + \frac{1}{2}y'(0)\frac{\lambda_1 + 1 - \lambda_2 - 1}{(\lambda_2 - \lambda_1)(\lambda_2 + 1)(\lambda_1 + 1)} +
$$

$$
+ \frac{3}{2}y(0)\frac{\lambda_2 - \lambda_1}{(\lambda_2 - \lambda_1)(\lambda_2 + 1)(\lambda_1 + 1)} + y(0)\frac{\lambda_1 - \lambda_2}{(\lambda_2 - \lambda_1)(\lambda_2 + 1)(\lambda_1 + 1)} -
$$

$$
- \frac{1}{2}y'(n)\frac{\lambda_2(\lambda_2 + 1)^{n-1} - \lambda_1(\lambda_1 + 1)^{n-1}}{\lambda_2 - \lambda_1} - \frac{1}{2}y'(n)\frac{(\lambda_2 + 1)^{n-1} - (\lambda_1 + 1)^{n-1}}{\lambda_2 - \lambda_1} +
$$

$$
+ \frac{3}{2}y(n)\frac{\lambda_2(\lambda_2 + 1)^{n-1} - \lambda_1(\lambda_1 + 1)^{n-1}}{\lambda_2 - \lambda_1} + y(n)\frac{(\lambda_2 + 1)^{n-1} - (\lambda_1 + 1)^{n-1}}{\lambda_2 - \lambda_1} =
$$

$$
= y(0)
$$

or equivalently

$$\frac{-1}{2}y'(0) + \frac{1}{2}y'(0) + \frac{3}{2}y(0) - y(0) - \frac{1}{2}y'(n)\frac{(\lambda_2+1)^n - (\lambda_1+1)^n}{\lambda_2 - \lambda_1} +$$

$$+ \frac{3}{2}y(n)\frac{\lambda_2(\lambda_2+1)^{n-1} - \lambda_1(\lambda_1+1)^{n-1}}{\lambda_2 - \lambda_1} + y(n)\frac{(\lambda_2+1)^{n-1} - (\lambda_1+1)^{n-1}}{\lambda_2 - \lambda_1} =$$

$$= y(0) \ .$$

$$(4.22)$$

From these expressions we conclude that

$$2\sqrt{3}y(0) + y'(n)\left[(\lambda_2+1)^n - (\lambda_1+1)^n\right] - 3y(n)\left[\lambda_2(\lambda_2+1)^{n-1} - \right.$$
$$\left. - \lambda_1(\lambda_1+1)^{n-1}\right] - 2y(n)\left[(\lambda_2+1)^{n-1} - (\lambda_1+1)^{n-1}\right] = 0 \ .$$

Now for the equation

$$y''(x) - 2y'(x) - 2y(x) = 0 \ . \qquad (4.23)$$

We consider the boundary condition

$$y'(n) - 2y(n) = 1(\lambda_2 - \lambda_1)y(0) - \left[\lambda_2(\lambda_2+1)^{n-1} - \lambda_1(\lambda_1+1)^{n-1}\right]y(n) = 0 \ .$$

$$(4.24)$$

If the general solution of the equation is in the form of

$$y(x) = c_1(1+\lambda_1)^x + c_2(1+\lambda_2)^x$$

then using the boundary conditions for c_1, c_2 we get the following system of equations

$$c_1\lambda_1(\lambda_1+1)^n + c_2\lambda_2(\lambda_2+1)^n - 2c_1(\lambda_1+1)^n - 2c_2(\lambda_2+1)^n = 1$$

$$(\lambda_2 - \lambda_1)[c_1 + c_2] - \left[\lambda_2(\lambda_2+1)^{n-1} - \lambda_1(\lambda_1+1)^{n-1}\right] \times$$

$$\times \left[c_1(\lambda_1+1)^n + c_2(\lambda_2+1)^n\right] = 0 \ .$$

By rearranging this system in the form of

$$\begin{cases} c_1 \left[\lambda_1(\lambda_1 + 1)^n - 2(\lambda_1 + 1)^n \right] + c_2 \left[\lambda_2(\lambda_2 + 1)^n - 2(\lambda_2 + 1)^n \right] = 1 \\ c_1 \left[(\lambda_2 - \lambda_1) - (\lambda_1 + 1)^n \left((\lambda_2 + 1)^{n-1}\lambda_2 - (\lambda_1 + 1)^{n-1}\lambda_1 \right) \right] + \\ +c_2 \left[(\lambda_2 - \lambda_1) - (\lambda_2 + 1)^n \left((\lambda_2 + 1)^{n-1}\lambda_2 - \lambda_1(\lambda_1 + 1)^{n-1} \right) \right] = 0 \ . \end{cases}$$

We can easily show that the main determinant, i.e., Δ is equal to zero, but one of the auxiliary determinants, e.g., Δ_1 is different from zero, in fact

$$\Delta_1 = \lambda_2 \left[(1 + \lambda_1) - (1 + \lambda_2)^{2n-1} \right] \neq 0\Gamma$$

This shows that the problem (4.23)-(4.24) has no solution.

4.5 The boundary value problem on an infinite domain

In this section we examine the following problem

$$y''(x) - 2y'(x) - 2y(x) = 0 \ , \qquad x \geq 0 \qquad\qquad (4.25)$$

$$y'(0) + \alpha y(0) = \beta \ , \qquad y(\infty) = 0 \ . \qquad\qquad (4.26)$$

Suppose that the general solution of (4.25) is the following

$$y(x) = c_1(1 + \lambda_1)^x + c_2(1 + \lambda_2)^x \qquad\qquad (4.27)$$

where according to the equalities

$$1 + \lambda_1 = 2 - \sqrt{3} \ , \quad 1 + \lambda_2 = 2 + \sqrt{3} \ .$$

We have the following inequalities

$$0 < 1 + \lambda_1 < 1 \ , \quad 3 < 1 + \lambda_2 < 4 \ .$$

Therefore by assuming $x \to \infty$ (on discrete values) then $(1 + \lambda_2)^x \to \infty$. From this it follows that $c_2 = 0$, so the general solution reduces to

$$y(x) = c_1(2 - \sqrt{3})^x \ .$$

Now to calculate c_1 we use the conditions (4.26), that is

$$c_1(1 - \sqrt{3}) + \alpha c_1 = \beta$$

or

$$c_1 = \frac{\beta}{1 - \sqrt{3} + \alpha} \ .$$

Therefore by assuming

$$1 - \sqrt{3} + \alpha \neq 0$$

for the solution we have

$$y(x) = \frac{\beta(2 - \sqrt{3})^x}{1 - \sqrt{3} + \alpha} \qquad x \geq 0 \ . \tag{4.28}$$

Now we discuss a non-homogeneous boundary value problem

$$y''(x) - 2y'(x) - 2y(x) = f(x) \qquad x \geq 0 \tag{4.29}$$

$$y(0) = y(\infty) = 0 \ . \tag{4.29_1}$$

As we have already observed, the general solution of the corresponding homogeneous equation of (4.29) is the following

$$y(x) = c_1(1 + \lambda_1)^x + c_2(1 + \lambda_2)^x$$

where

$$3 < 1 + \lambda_1 < 4 \ , \qquad 0 < 1 + \lambda_1 < 1 \ .$$

We are now seek for a bounded particular solution of the above non-homogeneous equation on an infinite domain. For this purpose, we use the method of variation of parameters, so suppose that the solution is

$$y(x) = \sum_{j=1}^{2} c_j(x)(1 + \lambda_j)^x \ . \tag{4.30}$$

By differentiating we get

$$y'(x) = \sum_{j=1}^{2} c_j'(x)(1 + \lambda_j)^{x+1} + \sum_{j=1}^{2} c_j(x)\lambda_j(1 + \lambda_j)^x$$

whence by assuming

$$\sum_{j=1}^{2} c_j'(x)(1 + \lambda_j)^{x+1} = 0 \tag{4.31}$$

we obtain the following

$$y'(x) = \sum_{j=1}^{2} c_j(x)\lambda_j(1 + \lambda_j)^x . \tag{4.32}$$

Finally if we differentiate (4.32) once more and substitute the resulting expression along with (4.30) and (4.32) into (4.29) we have

$$\sum_{j=1}^{2} c_j'(x)\lambda_j(1 + \lambda_j)^{x+1} + \sum_{j=1}^{2} c_j\lambda_j^2(1 + \lambda_j)^x -$$

$$- 2\sum_{j=1}^{2} c_j(x)\lambda_j(1 + \lambda_j)^x - 2\sum_{j=1}^{2} c_j(x)(1 + \lambda_j)^x = f(x)$$

or

$$\sum_{j=1}^{2} c_j'(x)\lambda_j(1 + \lambda_j)^{x+1} = f(x) . \tag{4.33}$$

If we think of the expressions (4.31) and (4.33) as an algebraic system, then we have

$$\begin{cases} (1 + \lambda_1)^{1+x}c_1'(x) + (1 + \lambda_2)^{1+x}c_2'(x) = 0 \\ (1 + \lambda_1)^{1+x}\lambda_1 c_1'(x) + (1 + \lambda_2)^{1+x}\lambda_2 c_2'(x) = f(x) \end{cases}$$

whence for $c_1'(x)$ and $c_2'(x)$ we get

$$c_1'(x) = -\frac{(1 + \lambda_1)^{-1-x}f(x)}{\lambda_2 - \lambda_1}$$

$$c_1'(x) = \frac{(1 + \lambda_2)^{-1-x}f(x)}{\lambda_2 - \lambda_1} .$$

By discrete integration from these expressions we have

$$c_1(x) = c_1 - \int_0^x \frac{(1+\lambda_1)^{-1}}{\lambda_2 - \lambda_1}(1+\lambda_1)^{-\xi}f(\xi)\Delta\xi$$

$$\int_x^\infty c_2'(\xi)d\xi = \int_x^\infty \frac{(1+\lambda_2)^{-1}}{\lambda_2 - \lambda_1}(1+\lambda_2)^{-\xi}f(\xi)\Delta\xi .$$

Therefore, for the general solution of (4.29) from (4.30) we get

$$y(x) = c_1(1+\lambda_1)^x - \int_0^x \frac{(1+\lambda_2)^{-1}}{\lambda_2 - \lambda_1}(1+\lambda_1)^{x-\xi}f(\xi)\Delta\xi+$$

$$+ c_2(1+\lambda_2)^x - \int_x^\infty \frac{(1+\lambda_2)^{-1}}{\lambda_2 - \lambda_1}(1+\lambda_2)^{x-\xi}f(\xi)\Delta\xi$$

or

$$y(x) = c_1(1+\lambda_1)^x + c_2(1+\lambda_2)^x + \int_0^\infty g(x-\xi)f(\xi) \tag{4.34}$$

where

$$g(x-\xi) = \begin{cases} -\frac{(1+\lambda_1)^{x-\xi}}{(1+\lambda_1)(\lambda_2-\lambda_1)} & 0 \le \xi < x \\ -\frac{(1+\lambda_2)^{x-\xi}}{(1+\lambda_2)(\lambda_2-\lambda_1)} & x \le \xi < \infty . \end{cases} \tag{4.35}$$

Finally by imposing the boundary conditions (4.29_1) we have

$$c_1 = - \int_0^\infty g(-\xi)f(\xi)\Delta\xi , \qquad c_2 = 0 .$$

Thus the solutions of (4.25_1)–(4.29) is

$$y(x) = \int_0^\infty G(x,\xi)f(\xi) \tag{4.36}$$

where

$$G(x,\xi) = g(x-\xi) - g(-\xi)(1+\lambda_1)^x . \tag{4.37}$$

Appendix A

2-Variable Discrete Analysis

[1] **Abstract.** This chapter deals with ten different problems. Except for the three sections; namely an unsolved problem about a cubic root of a complex number, symmetric derivative, and the spectral decomposition of a matrix by a projection matrices, all other sections are devoted to the discrete analysis of complex valued functions. Topics included are the discrete partial derivatives of complex valued functions, the Cauchy theorem, Laplace equation, Cauchy-Riemann discrete differential equations, complex integration and above all, the most important concept which brings all these together, is the discrete analog of analytic functions in complex analysis.

A.1 Discrete symmetric derivative and its application

In this section, let us suppose that the domain of functions is the set of $\left\{ \frac{k}{2} \mid k \in Z \right\}$ or its arbitrary subsets. We denote this set by $\frac{Z}{2}$. Then for any

---

function f defined on this set, we define its symmetric derivative as

$$f'(x) = f\left(x + \frac{1}{2}\right) - f\left(x - \frac{1}{2}\right). \tag{A.1}$$

This derivative can be used to derive some useful information about the functions which are symmetric around x. By calculating the second derivative we have

$$f''(x) = f(x+1) - f(x) - [f(x) - f(x-1)]$$
$$= f(x+1) - 2f(x) + f(x-1).$$

It is clear from this expression that if x is an integer, then the variables appeared in the second derivative are all integers. It can be easily seen that some properties of the one-directional derivative are also valid for this two-directional derivative.

a) The derivative of a constant function is equal to zero, that is

$$c' = f'(x) = f\left(x + \frac{1}{2}\right) - f\left(x - \frac{1}{2}\right) = c - c = 0.$$

b) The derivative of the identity function is equal to unity, that is

$$x' = f'(x) = f\left(x + \frac{1}{2}\right) - f\left(x - \frac{1}{2}\right)$$
$$= \left(x + \frac{1}{2}\right) - \left(x - \frac{1}{2}\right) = 1.$$

c) The derivative of a power function.

i) $(x^2)' = f\left(x + \frac{1}{2}\right) - f\left(x - \frac{1}{2}\right) = \left(x + \frac{1}{2}\right)^2 - \left(x - \frac{1}{2}\right)^2$

$$= x^2 + x + \frac{1}{4} - \left(x^2 - x + \frac{1}{4}\right)$$

$$= 2x.$$

ii) $(x^3)' = f\left(x+\dfrac{1}{2}\right) - f\left(x-\dfrac{1}{2}\right) = \left(x+\dfrac{1}{2}\right)^3 - \left(x-\dfrac{1}{2}\right)^3$

$\qquad = x^3 + \dfrac{3}{2}x^2 + \dfrac{3}{4}x + \dfrac{1}{8} - \left(x^3 - \dfrac{3}{2}x^2 + \dfrac{3}{4}x - \dfrac{1}{8}\right)$

$\qquad = 3x^2 + \dfrac{1}{4}$.

d) The derivative of the arrangement power function.

  i) $f(x) = x^{[2]} = x(x-1)$

$\qquad \left(x^{[2]}\right)' = \left(x+\dfrac{1}{2}\right)^{[2]} - \left(x-\dfrac{1}{2}\right)^{[2]}$

$\qquad\qquad = \left(x+\dfrac{1}{2}\right)\left(x-\dfrac{1}{2}\right) - \left(x-\dfrac{1}{2}\right)\left(x-\dfrac{3}{2}\right)$

$\qquad\qquad = \left(x-\dfrac{1}{2}\right)\left[x+\dfrac{1}{2} - \left(x-\dfrac{3}{2}\right)\right] = 2\left(x-\dfrac{1}{2}\right)$.

  ii) $f(x) = x^{[3]} = x(x-1)(x-2)$

$\qquad \left(x^{[3]}\right)' = \left(x+\dfrac{1}{2}\right)^{[3]} - \left(x-\dfrac{1}{2}\right)^{[3]}$

$\qquad\qquad = \left(x+\dfrac{1}{2}\right)\left(x-\dfrac{1}{2}\right)\left(x-\dfrac{3}{2}\right) - \left(x-\dfrac{1}{2}\right)\left(x-\dfrac{3}{2}\right)\left(x-\dfrac{5}{2}\right)$

$\qquad\qquad = \left(x-\dfrac{1}{2}\right)\left(x-\dfrac{3}{2}\right)\left[x+\dfrac{1}{2} - \left(x-\dfrac{5}{2}\right)\right]$

$\qquad\qquad = 3\left(x-\dfrac{1}{2}\right)^{[2]}$.

In general

$$\left(x^{[n]}\right)' = n\left(x-\dfrac{1}{2}\right)^{[n-1]} .$$

e) The derivative of the arrangement power function with step $\frac{1}{2}$.

Assume that

  i) $x^{[2]}_{1/2} = x\left(x - \frac{1}{2}\right)$

ii) $x_{1/2}^{[n]} = x \left(x - \frac{1}{2} \right) \left(x - \frac{2}{2} \right) \cdots \left(x - \frac{n-1}{2} \right)$

Then we have

i) $\left(x_{1/2}^{[2]} \right)' = \left(x + \frac{1}{2} \right)_{1/2}^{[2]} - \left(x - \frac{1}{2} \right)_{1/2}^{[2]}$

$= \left(x + \frac{1}{2} \right) x - \left(x - \frac{1}{2} \right) (x - 1)$

$= 2x - \frac{1}{2} \ .$

ii) $\left(x_{1/2}^{[3]} \right)' = \left(x + \frac{1}{2} \right)_{1/2}^{[3]} - \left(x - \frac{1}{2} \right)_{1/2}^{[3]}$

$= \left(x + \frac{1}{2} \right) x \left(x - \frac{1}{2} \right) - \left(x - \frac{1}{2} \right) (x - 1) \left(x - \frac{3}{2} \right)$

$= 3 \left(x - \frac{1}{2} \right)^{[2]} \ .$

f) The function which is invariant under the symmetric derivative.

Suppose that f is the required function, then

$$f'(x) = f \left(x + \frac{1}{2} \right) - f \left(x - \frac{1}{2} \right) = f(x)$$

or equivalently

$$f \left(x + \frac{1}{2} \right) = f(x) + f \left(x - \frac{1}{2} \right)$$

and this is just a function which satisfies the Fibonacci relation with step $\frac{1}{2}$.

Finally the definition of the symmetric integral is exactly the same as the discrete integral except for the step which is $\frac{1}{2}$ instead. We point out that its properties are almost similar but because of some complex expressions in the derivative of the product of two functions, integration by parts will be a little bit more difficult.

A.2   Function of two variables, partial derivative and discrete analytic function

Suppose that the function $f(x, y)$ with $z \in Z$, $y \in Z$ is given. We have the following definition.

Definition. By the discrete partial derivative of $f(x, y)$ with respect to x we mean the following

$$\frac{\partial f(x, y)}{\partial x} = f(x + 1, y) - f(x, y) \ . \tag{A.2}$$

Similarly for the variable y we have

$$\frac{\partial f(x, y)}{\partial y} = f(x, y + 1) - f(x, y) \ .$$

Now we are going to examine the functions which satisfy some kind of analytic condition, these functions will be simply called the discrete analytic functions. To this end, suppose that $Z = x + iy$, where $x, y \in Z$ and $i^2 = -1$. Also suppose that $f : Z^2 \to Z^2$ is defined by

$$f(Z) = f(x + iy) = U(x + iy) + iV(x + iy) \tag{A.3}$$

where U and V are real and imaginary parts of f respectively. As usual, we denote these functions by

$$U(x + iy) = \operatorname{Re} f(x + iy) \ , \quad V(x + iy) = \operatorname{Im} f(x + iy) \ .$$

The following definition seems to be natural.

Definition. *Analytic function.* If the function $f : Z^2 \to Z^2$ satisfies in the equations of

$$\begin{cases} \frac{\partial U(x,y)}{\partial x} = \frac{\partial V(x,y)}{\partial y} \\ \frac{\partial U(x,y)}{\partial y} = -\frac{\partial V(x,y)}{\partial x} \end{cases} \tag{A.4}$$

then f is called an analytic function in two discrete variables x and y.

Examples of analytic functions.   Define

$$Z^{[1]} = (x + iy)$$

$$Z^{[2]} = (x + iy)^{[2]} = x^{[2]} - y^{[2]} + 2ixy$$

$$Z^{[3]} = (x + iy)^{[3]} = x^{[3]} + 3x^{[2]}yi - 3xy^{[2]} - iy^{[3]}$$

$$\vdots$$

$$Z^{[n]} = (x + iy)^{[n]} = \sum_{k=0}^{n} C_n^k x^{[n-k]} y^{[k]} i^k$$

$$= \left(x^{[n]} - C_n^2 x^{[n-2]} y^{[2]} + C_n^4 x^{[n-4]} y^{[4]} - \cdots \right)$$

$$+ i \left(C_n^1 x^{[n-1]} y - C_n^3 x^{[n-3]} y^{[3]} + \cdots \right) .$$

(A.5)

Then $Z^{[n]}$ is an analytic function for each $n \in \mathbb{N}$, in fact we have

$$U(x,y) = x^{[n]} - C_n^2 x^{[n-2]} y^{[2]} + C_n^4 x^{[n-4]} y^{[4]} - \cdots$$

$$V(x,y) = C_n^1 x^{[n-1]} y - C_n^3 x^{[n-3]} y^{[3]} + \cdots$$

(A.6)

whence

$$\frac{\partial U(x,y)}{\partial x} = n x^{[n-1]} - \frac{n(n-1)}{2}(n-2)x^{[n-3]} y^{[2]} +$$

$$+ \frac{n(n-1)(n-2)(n-3)}{4!}(n-4)x^{[n-5]} y^{[4]} - \cdots$$

$$\frac{\partial U(x,y)}{\partial y} = -\frac{n(n-1)}{2} x^{[n-2]} 2y + \frac{n(n-1)(n-2)(n-3)}{4!} x^{[n-4]} 4y^{[3]} - \cdots$$

$$\frac{\partial V(x,y)}{\partial x} = n(n-1)x^{[n-2]} y - \frac{n(n-1)(n-2)(n-3)}{3!} x^{[n-4]} y^{[3]} + \cdots$$

$$\frac{\partial V(x,y)}{\partial y} = n x^{[n-1]} - \frac{n(n-1)(n-2)}{3!} x^{[n-3]} 3y^{[2]} + \cdots .$$

It follows that

$$\begin{cases} \frac{\partial U}{\partial x} = \frac{\partial V}{\partial y} \\ \frac{\partial U}{\partial y} = -\frac{\partial V}{\partial x} . \end{cases}$$

i.e., the function Z^n, $n \in \mathbb{N}$ is an analytic function with respect to the complex discrete variable $Z = x + iy$.

It is clear from the above facts that the complex polynomial function

$$f(Z) = \sum_{k=0}^{n} a_k Z^{[n]} \tag{A.7}$$

is an analytic functions where a_k $(k = 0, 1, \ldots, n)$ are arbitrary real constants. It is easily seen that if the coefficients a_k are complex numbers, the function $f(Z)$ still remains analytic.

Next we consider a function which behaves like the exponential function e^Z in complex analysis. We define this function as

$$f(x + iy) = 2^x (1 + i)^y \qquad x, y \in Z \ .$$

Let us suppose that the real functions $u(x)$ and $v(x)$ are the real and imaginary parts of f respectively, then

$$2^x (1 + i)^y = U(x, y) + iV(x, y) \ .$$

First of all, we know from Chapter 3 that

$$2^x = \sum_{k=0}^{\infty} \frac{x^{[k]}}{k!}$$

and

$$(1 + i)^y = \cos[y] = i \sin[y]$$

where

$$\cos[y] = \frac{(1 + i)^y + (1 - i)^y}{2} \ , \qquad \sin[y] = \frac{(1 + i)^x - (1 - i)^x}{2i} \ .$$

Then for the function $f(x + iy) = 2^x (1 + i)^y$ we have

$$2^x (1 + i)^y = 2^x \cos[y] + i2^x \sin[y] \ .$$

That is,

$$U(x,y) = 2^x \cos[y] \;, \qquad V(x,y) = 2^x \sin[y]$$

whence

$$\frac{\partial U}{\partial x} = 2^x \cos[y] \qquad \frac{\partial V}{\partial x} = 2^x \cos[y]$$

$$\frac{\partial U}{\partial y} = -2^x \sin[y] \qquad \frac{\partial V}{\partial x} = 2^x \sin[y] \;.$$

¿From these discrete Cauchy-Riemann equations we see that the function f is an analytic function. This function is called the discrete complex exponential function.

A.3 Discrete transformation of the second type

We discussed the discrete transformation in Chapter 2, and utilized it for the periodic finite sequence scheme or non-periodic finite sequence scheme. In this article, we wish to examine the non-periodic infinite sequences. To do this, consider the binomial equation of

$$x^{\sqrt{2}} - 1 = 0 \qquad\qquad (A.8)$$

where according to $e^{2k\pi i} = 1$ we have

$$x_k = \varepsilon^k \qquad k \in Z \qquad\qquad (A.9)$$

$$= e^{\sqrt{2}\pi i k} \qquad\qquad (A.10)$$

By assuming the sequence $(a_k)_{k \in Z}$, we correspond the function $A(t)$ as

$$A(t) = \sum_{k \in Z} \varepsilon^{-kt} a_k \;. \qquad\qquad (A.11)$$

We have the following expressions

$$\int_0^{\sqrt{2}} \varepsilon^{tm} A(t)dt = \int_0^{\sqrt{2}} \varepsilon^{tm} dt \sum_{k\in Z} \varepsilon^{-kt} a_k$$

$$= \sum_{k\in Z} a_k \int_0^{\sqrt{2}} \varepsilon^{t(m-k)} dt$$

$$= \sum_{k\in Z} a_k \left\{ \begin{array}{ll} \dfrac{e^{\sqrt{2}\pi i(m-k)t}}{\sqrt{2}\pi i(m-k)} \Big|_0^{\sqrt{2}} & m \neq k \\[2mm] \sqrt{2} & m = k \end{array} \right\} = \sqrt{2}a_m$$

i.e.,

$$a_m = \frac{1}{\sqrt{2}} \int_0^{\sqrt{2}} \varepsilon^{tm} A(t)dt \ . \tag{A.12}$$

In this way, (A.11)–(A.12) is the transformation for the non-periodic infinite sequence scheme. As we observe, this transformation is the Fourier series in mathematical analysis, in fact we established a one-to-one correspondence between infinite sequences and the functions $A(t)$ where the terms of the sequence are the coefficients of the corresponding Fourier series.

Remark 1.   If for the non-periodic infinite sequence $(a_k)_{k\in Z}$ we correspond

$$A_p = \sum_{k\in Z} \varepsilon^{-kp} a_k \qquad p \in Z$$

then

$$\sum_{p\in Z} \varepsilon^{pm} A_p = \sum_{p\in Z} \varepsilon^{pm} \sum_{k\in Z} \varepsilon^{-kp} a_k$$

$$= \sum_{k\in Z} a_k \sum_{p\in Z} \varepsilon^{p(m-k)}$$

where for $m \neq k$ we have

$$\sum_{p \in Z} \varepsilon^{p(m-k)} = \sum_{p \in Z} e^{\sqrt{2}\pi i p(m-k)}$$

$$= 2\pi \sum_{p \in Z} \delta \left[\sqrt{2}\pi (m-k) - 2\pi p \right]$$

$$= 2\pi \sum_{p \in Z} \delta \left\{ \pi \left[\sqrt{2}(m-k) - 2p \right] \right\} = 0$$

since

$$m, k, p \in Z, we have \qquad \sqrt{2}(m-k) - 2p \neq 0 .$$

Finally, since for $m = k$ we have

$$\sum_{p \in Z} \varepsilon^{p(m-k)} = \infty$$

the inverse transformation can not be defined.

Remark 2. If we wish to apply the above transformation for the solution of system of equations

$$\sum_{j=0}^{m} \alpha_j a_{j+p} = b_p , \qquad p \in Z$$

then the condition

$$\sum_{j=0}^{m} \alpha_j \varepsilon^{jp} \neq 0 \qquad p \in Z$$

seems to be necessary. This condition for $p = 0$ is

$$\sum_{j=0}^{m} \alpha_j \neq 0 .$$

This means that for the solution of the equation

$$y^{(m)}(x) = f(x)$$

which can be readily solved, the above transformation is not applicable.

A.4   Fundamental solutions

Consider the Cauchy differential equation with discrete variables (x_1, x_2)

$$\frac{\partial U(x)}{\partial x_2} + i\frac{\partial U(x)}{\partial x_1} = \phi(x) \ . \tag{A.13}$$

We apply the discrete transformation of the first type (which is defined in Chapter 2) to this problem, i.e., by using (2.5) we seek for a discrete function $U(x) = U(x_1, x_2)$ in the form

$$U(x) = \frac{1}{n^2} \sum_{\alpha_1=0}^{n-1} \sum_{\alpha_2=0}^{n-1} \varepsilon^{x_1\alpha_1 + x_2\alpha_2} \widetilde{U}(a) \ . \tag{A.14}$$

Similarly for the function $\phi(x)$ we have the expression

$$\phi(x) = \frac{1}{n^2} \sum_{\alpha_1=0}^{n-1} \sum_{\alpha_2=0}^{n-1} \varepsilon^{x_1\alpha_1 + x_2\alpha_2} \tilde{\phi}(\alpha) \tag{A.15}$$

and the transformation

$$\tilde{\phi}(\alpha) \sum_{\xi_1=0}^{n-1} \sum_{\xi_2=0}^{n-1} \varepsilon^{-\alpha_1\xi_1 - \alpha_2\xi_2} \phi(\xi) \ . \tag{A.16}$$

By substituting the expressions (A.14) and (A.15) in (A.13) and assuming

$$[(\varepsilon^{\alpha_2} - 1) + i(\varepsilon^{\alpha_1} - 1)]\widetilde{U}(\alpha) = \tilde{\phi}(\alpha)$$

or equivalently

$$(\varepsilon^{\alpha_2} - 1) + i(\varepsilon^{\alpha_1} - 1) \neq 0 \tag{A.17}$$

then we have

$$\widetilde{U}(\alpha) = \frac{\tilde{\phi}(\alpha)}{(\varepsilon^{\alpha_2} - 1) + i(\varepsilon^{\alpha_1} - 1)} \ . \tag{A.18}$$

Finally, by substituting (A.18) in (A.10) and using (A.12) we get

$$U(x) = \frac{1}{n^2} \sum_{\alpha_1=0}^{n-1} \sum_{\alpha_2=0}^{n-1} \varepsilon^{\alpha_1 x_1 + \alpha_2 x_2} \frac{1}{(\varepsilon^{\alpha_2} - 1) + i(\varepsilon^{\alpha_1} - 1)} \times$$

$$\times \sum_{\xi_1=0}^{n-1} \sum_{\xi_2=0}^{n-1} \varepsilon^{-\alpha_1\xi_1 - \alpha_2\xi_2} \phi(\xi) \ .$$

In this way, we find the fundamental solution of (A.13) as

$$U(x - \xi) = \frac{1}{n^2} \sum_{\alpha_1=0}^{n-1} \sum_{\alpha_2=0}^{n-1} \frac{\varepsilon^{\alpha_1(x_1-\xi_1)+\alpha_2(x_2-\xi_2)}}{(\varepsilon^{\alpha_1} - 1) + i(\varepsilon^{\alpha_2} - 1)} \cdot \qquad (A.19)$$

We observe from the expression (A.19) that if α_1 and α_2 vanish simultaneously, then the denominator also vanishes, it follows that the fundamental solution (A.19) under the condition (A.17) fulfills our purpose. For this reason, if we wish the function $\widetilde{U}(\alpha)$ to be meaningful in (A.18), the equality (for $\alpha_1 = \alpha_2 = 0$) $\tilde{\phi}(0) = 0$ must hold. Having said that, from the expression (A.16) we see that the equality

$$\sum_{\xi_1=0}^{n-1} \sum_{\xi_2=0}^{n-1} \phi(\xi) = 0 \qquad (A.20)$$

holds. This contradicts the fact that the function being arbitrary.

Remark. This property also occurs for the continuous functions. Therefore, in order to solve the Cauchy-Riemann equation via the Fourier transformation, we have

$$\frac{\partial U(x)}{\partial x_2} = U(x_1, x_2 + 1) - U(x_1, x_2)$$

or

$$\frac{\partial U(x)}{\partial x_1} = U(x_1 + 1, x_2) - U(x_1, x_2) \ .$$

Taking into account these expressions, for the fundamental solution of (A.13) we first apply the discrete transformation of the second type with respect to one of the variables and then solve the resulting ordinary differential equation with respect to the other variable. For this reason, we substitute

$$U(x) = \frac{1}{\sqrt{2}} \int_0^{\sqrt{2}} \varepsilon^{tx_1} V(t, x_2) dt \qquad\qquad (A.21)$$

$$\phi(x) = \frac{1}{\sqrt{2}} \int_0^{\sqrt{2}} \varepsilon^{tx_1} \Psi(t, x_2) dt \qquad\qquad (A.22)$$

$$\Psi(t, x_2) = \sum_{\xi_1 \in Z} \varepsilon^{-\xi_1 t} \phi(\xi_1, x_2) \qquad\qquad (A.23)$$

in (A.13) to obtain

$$\frac{\partial V}{\partial x_2} + i(\varepsilon^t - 1)V = \Psi \qquad\qquad (A.24)$$

where $t \in (0, \sqrt{2})$ is a parameter and $\varepsilon = e^{\sqrt{2}\pi i}$.

The general solution for the corresponding homogeneous equation of (A.24) is

$$V(t, x_2) = c \left[1 - i(\varepsilon^t - 1)\right]^{x_2} .$$

Now by using the method of the variation of parameters we seek for the solution of (A.24) as

$$V(t, x_2) = \left[1 + i(1 - \varepsilon^t)\right]^{x_2} c(x_2) . \qquad\qquad (A.25)$$

By substituting (A.25) in (A.24) we have

$$i(1 - \varepsilon^t) \left[1 + i(1 - \varepsilon^t)\right]^{x_2} c'(x_2) +$$
$$+ i(1 - \varepsilon^t) \left[1 + i(1 - \varepsilon^t)\right]^{x_2} c(x_2) +$$
$$+ \left[1 + i(1 - \varepsilon^t)\right]^{x_2} c'(x_2) - i(1 - \varepsilon^t) \left[1 + i(1 - \varepsilon^t)\right]^{x_2} c(x_2) =$$
$$= \Psi(t, x_2)$$

or equivalently

$$c'(x_2) = \Psi(t, x_2) \left[1 + i(1 - \varepsilon^t)\right]^{-x_2 - 1} .$$

By integration we get

$$c(x_2) = c + \sum_{\xi_2 \leq x_2} \Psi(t, \xi_2) \left[1 + i(1 - \varepsilon^t)\right]^{-\xi_2} .$$

If we substitute this expression in (A.25) (assuming $c = 0$) for the particular solution of (A.24) we obtain

$$V(t, x_2) = \sum_{\xi_2 \in Z} \Psi(t, \xi_2) \left[1 + i(1 - \varepsilon^t)\right]^{x_2 - \xi_2} \theta(x_2 - \xi_2) \qquad (A.26)$$

where

$$\theta(x_2 - \xi_2) = \begin{cases} 1 & x_2 \geq \xi_2 \\ 0 & x_2 < \xi_2 . \end{cases} \qquad (A.27)$$

Finally by considering (A.23), from (A.21) we have

$$U(x) = \frac{1}{\sqrt{2}} \int_0^{\sqrt{2}} \varepsilon^{tx_1} t d \sum_{\xi_2 \in Z} \left[1 + i(1 - \varepsilon^t)\right]^{x_2 - \xi_2} \times$$
$$\times \theta(x_2 - \xi_2) \sum_{\xi_1 \in Z} \varepsilon^{-\xi_1 t} \phi(\xi) .$$

This in turn gives the following fundamental solution

$$V(x - \xi) = \frac{\theta(x_2 - \xi_2)}{\sqrt{2}} \int_0^{\sqrt{2}} \varepsilon^{t(x_1 - \xi_1)} \left[1 + i(1 - \varepsilon^t)\right]^{x_2 - \xi_2} dt . \qquad (A.28)$$

Remark. In mathematical physics courses, seeking the fundamental solution of the Cauchy-Riemann equation via Fourier transformation gives rise to a integral with a singular point. This difficulty is removed from the fundamental solution in (A.28). If we wish to apply this transformation with respect to both variables, then the appearance of the above singularity is unavoidable.

A.5 The fundamental solution for adjoint of the discrete Laplace equation

Suppose that the equation

$$\Delta U_{ij} = U_{(i+2)j} + U_{i(j+2)} - 2U_{(i+1)j} - 2U_{i(j+1)} + 2U_{ij} = 0$$

$$i = 0, 1, \ldots, n \ , \quad j = 1, 2, \ldots, m \tag{A.29}$$

is given. Consider the inner product

$$(\Delta U, V) = \sum_{i=0}^{n} \sum_{j=0}^{m} \Delta U_{ij} V_{ij}$$

$$= \sum_{i=0}^{n} \sum_{j=0}^{m} \left[U_{(i+2)j} + U_{i(j+2)} - 2U_{(i+1)j} - 2U_{i(j+1)} + 2U_{ij} \right] V_{ij}$$

$$= \sum_{i=2}^{n+2} \sum_{j=0}^{m} U_{ij} V_{(i-2)j} + \sum_{i=0}^{n} \sum_{j=2}^{m+2} U_{ij} V_{i(j-2)} -$$

$$- 2 \sum_{i=1}^{n+1} \sum_{j=0}^{m} U_{ij} V_{(i-1)j} - 2 \sum_{i=0}^{n} \sum_{j=1}^{m+1} U_{ij} V_{i(j-1)} +$$

$$+ 2 \sum_{i=0}^{n} \sum_{j=0}^{m} U_{ij} V_{ij}$$

$$= \sum_{i=0}^{n} \sum_{j=0}^{m} U_{ij} \left(V_{(i-2)j} + V_{i(j-2)} - 2V_{(i-1)j} - 2V_{i(j-1)} + 2V_{ij} \right) +$$

$$+ \sum_{j=0}^{m} \left(U_{(n+1)j} V_{(n-1)j} + U_{(n+2)j} V_{nj} - U_{0j} V_{-2j} - U_{1j} V_{-1j} \right) +$$

$$+ \sum_{i=0}^{n} \left(U_{i(m+1)} V_{i(m-1)} + U_{i(m+2)} V_{im} - U_{i0} V_{i(-2)} - U_{i1} V_{i(-1)} \right) -$$

$$- 2 \sum_{j=0}^{m} \left(U_{(n+1)j} V_{nj} - U_{0j} V_{-1j} \right) -$$

$$- 2 \sum_{i=0}^{n} \left(U_{i(m+1)} V_{im} - U_{i0} V_{i(-1)} \right) \ .$$

Denoting the expression inside the bracket of the double summation by $\Delta^* V_{ij}$ and the remaining sums by $\{**\}$, then we obtain the discrete analog of Lagrange formula as follows

$$(\Delta U, V) = (U, \Delta^* V) + \{**\} \qquad \Delta^* V = (\Delta^* V_{ij})_{ij} \qquad (\text{A.30})$$

where

$$\Delta^* V_{ij} = V_{(i-2)j} + V_{i(j-2)} - 2V_{(i-1)j} - 2V_{i(j-1)} + 2V_{ij} \qquad (\text{A.31})$$

is the adjoint of the discrete Laplace equation.

In this stage, we are going to construct the fundamental solution of the equation

$$\Delta^* V_{ij} = g_{ij} \qquad i, j \in Z . \qquad (\text{A.32})$$

For this purpose, as we did for the discrete Cauchy-Riemann equation, we first apply the discrete transformation of the second type with respect to the second variable j to obtain

$$V_{ij} = \frac{1}{\sqrt{2}} \int_0^{\sqrt{2}} V_i(\eta) \varepsilon^{\eta j} d\eta \qquad (\text{A.33})$$

$$g_{ij} = \frac{1}{\sqrt{2}} \int_0^{\sqrt{2}} G_i(\eta) \varepsilon^{\eta j} d\eta \qquad (\text{A.34})$$

$$G_i(\eta) = \sum_{q \in Z} \varepsilon^{-\eta q} g_{iq} \qquad (\text{A.35})$$

where $\varepsilon = e^{i\pi\sqrt{2}}$.

By substituting (A.33) and (A.34) in (A.32), we get

$$V_i(\eta) - 2V_{i-1}(\eta) + \left(\varepsilon^{2\eta} - 2\varepsilon^{-\eta} + 2 \right) V_i(\eta) = G_i(\eta)$$

or equivalently

$$V_{i-2}(\eta) - 2V_{i-1}(\eta) + \left[(\varepsilon^{-\eta} - 1)^2 + 1 \right] V_i(\eta) = G_i(\eta) . \qquad (\text{A.36})$$

Now suppose that the particular solution of the corresponding homogeneous differential equation of (A.36) is in the form

$$V_i(\eta) = \lambda^i .$$

This gives the characteristic equation of

$$(\lambda^{-1})^2 - 2\lambda^{-1} + (\varepsilon^{-\eta} - 1)^2 + 1 = 0 .$$

Since the roots are

$$\lambda^{-1} = 1 \pm \sqrt{-1}(\varepsilon^{-\eta} - 1)$$

so the general solution of the homogeneous equation is

$$V_i(\eta) = C_1 \left[1 - \sqrt{-1}(\varepsilon^{-\eta} - 1)\right]^{-i} + C_{2i} \left[1 + \sqrt{-1}(\varepsilon^{-\eta} - 1)\right]^{-i} .$$

Using the method of the variation of parameters we have

$$V_i(\eta) = C_{1i} \left[1 - \sqrt{-1}(\varepsilon^{-\eta} - 1)\right]^{-i} + C_{2i} \left[1 + \sqrt{-1}(\varepsilon^{-\eta} - 1)\right]^{-i} . \qquad (A.37)$$

By calculating the first derivative we see that

$$V_{i-1}(\eta) = \left(C_{1(i-1)}C_{1i}\right) \left[1 - \sqrt{-1}(\varepsilon^{-\eta} - 1)\right]^{i+1} +$$

$$+ \left(C_{2(i-1)} - C_{2i}\right) \left[1 + \sqrt{-1}(\varepsilon^{-\eta} - 1)\right]^{-i+1} +$$

$$+ C_{1i} \left[1 - \sqrt{-1}(\varepsilon^{-\eta} - 1)\right]^{-i+1} C_{2i} \left[1 + \sqrt{-1}(\varepsilon^{-\eta} - 1)\right]^{-i+1} .$$

Assuming

$$\left(C_{1(i-1)} - C_{1i}\right) \left[1 - \sqrt{-1}(\varepsilon^{-\eta} - 1)\right]^{-i+1} +$$

$$+ \left(C_{2(i-1)} - C_{2i}\right) \left[1 + \sqrt{-1}(\varepsilon^{-\eta} - 1)\right]^{-i+1} = 0 . \qquad (A.38)$$

By substituting the remaining part of $V_{i-1}(\eta)$ along with (A.37) in (A.36) we get

$$
\begin{aligned}
\left(C_{1(i-1)} - C_{1i}\right) &\left[1 - \sqrt{-1}(\varepsilon^{-\eta} - 1)\right]^{-i+2} + \\
+ \left(C_{2(i-1)} - C_{2i}\right) &\left[1 + \sqrt{-1}(\varepsilon^{-\eta} - 1)\right]^{-i+2} + \\
+ C_{1i}\Big\{ &\left[1 - \sqrt{-1}(\varepsilon^{-\eta} - 1)\right]^2 - 2\left[1 - \sqrt{-1}(\varepsilon^{-\eta} - 1)\right] + \\
&+ (\varepsilon^{-\eta} - 1)^2\Big\} \left[1 - \sqrt{-1}(\varepsilon^{-\eta} - 1)\right]^{-i} + \\
+ C_{2i}\Big\{ &\left[1 + \sqrt{-1}(\varepsilon^{-\eta} - 1)\right]^2 - 2\left[1 + \sqrt{-1}(\varepsilon^{-\eta} - 1)\right] + \\
&+ (\varepsilon^{-\eta} - 1)^2 + 1\Big\} \left[1 + \sqrt{-1}(\varepsilon^{-\eta} - 1)\right]^{-i} \\
= G_i(\eta)
\end{aligned}
$$

where since the expression inside the right bracket is the characteristic equation, it is equal to zero. Therefore the remaining expressions along with (A.38) give rise to the following system of equations

$$
\begin{cases}
\left(C_{1(i-1)} - C_{1i}\right)\left[1 - \sqrt{-1}(\varepsilon^{-\eta} - 1)\right]^{-i+1} + \\
+ \left(C_{2(i-1)} - C_{2i}\right)\left[1 + \sqrt{-1}(\varepsilon^{-\eta} - 1)\right]^{-i+1} = 0 \\
\left(C_{1(i-1)} - C_{1i}\right)\left[1 - \sqrt{-1}(\varepsilon^{-\eta} - 1)\right]^{-i+2} + \\
+ \left(C_{2(i-1)} - C_{2i}\right)\left[1 + \sqrt{-1}(\varepsilon^{-\eta} - 1)\right]^{-i+2} = G_i(\eta) \; .
\end{cases}
$$

This system is in turn equivalent to

$$
\begin{cases}
C_{1(i-1)} - C_{1i} &= -\dfrac{\left[1 - \sqrt{-1}(\varepsilon^{-\eta}-1)\right]^{i-1}}{2\sqrt{-1}(\varepsilon^{-\eta}-1)}G_i(\eta) \\[2mm]
C_{2(i-1)} - C_{2i} &= \dfrac{\left[1 + \sqrt{-1}(\varepsilon^{-\eta}-1)\right]^{i-1}}{2\sqrt{-1}(\varepsilon^{-\eta}-1)}G_i(\eta) \; .
\end{cases}
$$

By integration we find out

$$
\begin{cases}
C_{1i} = C_1(-\infty) + \sum_{k=-\infty}^{i}\dfrac{\left[1 - \sqrt{-1}(\varepsilon^{-\eta}-1)\right]^{k-1}}{2\sqrt{-1}(\varepsilon^{-\eta}-1)}G_k(\eta) & i \in Z \\[3mm]
C_{2i} = C_2(-\infty) - \sum_{k=-\infty}^{i}\dfrac{\left[1 + \sqrt{-1}(\varepsilon^{-\eta}-1)\right]^{k-1}}{2\sqrt{-1}(\varepsilon^{-\eta}-1)}G_k(\eta) & i \in Z \; .
\end{cases}
$$

By taking $C_1 = C_2 = 0$ in the above expressions and substituting the resulting expression in (A.37), the particular solution of (A.36) will be

$$(V_i) = \sum_{k=-\infty}^{i} \frac{\left[1 - \sqrt{-1}(\varepsilon^{-\eta} - 1)\right]^{k-i-1} - \left[1 + \sqrt{-1}(\varepsilon^{-\eta} - 1)\right]^{k-i-1}}{2\sqrt{-1}(\varepsilon^{-\eta} - 1)} G_k(\eta)$$

$$(A.38')$$

Finally by substituting $(A.38')$ and (A.35) in (A.33) the solution for the equation of (A.32) is

$$V_{ij} = \frac{1}{\sqrt{2}} \int_0^{\sqrt{2}} \varepsilon^{\eta j} d\eta \times$$

$$\times \sum_{k \in Z} \Theta(i - k) \frac{\left[1 - \sqrt{-1}(\varepsilon^{-\eta} - 1)\right]^{k-i-1} - \left[1 + \sqrt{-1}(\varepsilon^{-\eta} - 1)\right]^{k-i-1}}{2\sqrt{-1}(\varepsilon^{-\eta} - 1)} \times$$

$$\times \sum_{p \in Z} \varepsilon^{-\eta q} g_{kq} .$$

It follows from this that the fundamental solution is

$$V_{i-k,j-q} = \frac{\Theta(i - k)}{2\sqrt{2}\sqrt{-1}} \int_0^{\sqrt{2}} \frac{\varepsilon^{\eta(j-q)}}{\varepsilon^{-\eta} - 1} \times$$

$$\times \left\{ \left[1 - \sqrt{-1}(\varepsilon^{-\eta} - 1)\right]^{k-i-1} - \left[1 + \sqrt{-1}(\varepsilon^{-\eta} - 1)\right]^{k-i-1} \right\} d\eta \quad (A.39)$$

where

$$\Theta(i - k) = \begin{cases} 1 & k \leq i \\ 0 & k > i \end{cases} \quad (A.40)$$

is the discrete analog of the identity Heaviside function.

A.6   Boundary value problem for the discrete Laplace equation

If in discrete Lagrange formula obtained in the equation (A.30) instead of U_{ij} and V_{ij} we substitute the solution of the equation (A.29) and the fundamental solution of (A.32) i.e., the expression (A.39) respectively, then we have

$$
2\sum_{j=0}^{m} \left(U_{n+1,j}V_{n-k,j-q} - U_{0j}V_{-1-k,j-q}\right) +
$$

$$
+ 2\sum_{i=0}^{n} \left(U_{i(m+1)}V_{i-k,m-j} - U_{10}V_{i-k-1-q}\right) -
$$

$$
- \sum_{j=0}^{m} \Big(U_{n+1,j}V_{n-k-1,j-q} + U_{n+2,j}V_{n-k,j-q} -
$$

$$
- U_{0j}V_{-2-k,j-q} - U_{1j}V_{-1-k,j-q}\Big) -
$$

$$
- \sum_{i=0}^{n} \Big(U_{i(m+1)}V_{i-k,m-1-q} + U_{i(m+2)}V_{i-k,m-q} -
$$

$$
- U_{i0}V_{i-k,-2-q} - U_{i1}V_{i-k,-1-q}\Big) =
$$

$$
= \sum_{i=0}^{n}\sum_{j=0}^{m} U_{ij}\Big(V_{i-2-k,j-q} + V_{i-k,j-2-q} -
$$

$$
- 2V_{i-1-k,j-q} - 2V_{i-k,j-1-q} + 2V_{i-k,j-q}\Big) .
$$

(A.41)

Now we calculate the expression inside the bracket

$$V_{i-2-k,j-q} + V_{i-k,j-2-q} - 2V_{i-1-k,j-q}-$$

$$- 2V_{i-k,j-1-q} + 2V_{i-k,j-q} =$$

$$= \frac{\Theta(i-2-k)}{2\sqrt{2}\sqrt{-1}} \int_0^{\sqrt{2}} \frac{\varepsilon^{\eta(j-q)}}{\varepsilon^{-\eta}-1} \times$$

$$\times \left\{ \left[1 - \sqrt{-1}(\varepsilon^{-\eta}-1)\right]^{k-i+1} - \left[1 + \sqrt{-1}(\varepsilon^{-\eta}-1)\right]^{k-i+1} \right\} d\eta +$$

$$+ \frac{\Theta(i-k)}{2\sqrt{2}\sqrt{-1}} \int_0^{\sqrt{2}} \frac{\varepsilon^{\eta(j-2-q)}}{\varepsilon^{-\eta}-1} \times$$

$$\times \left\{ \left[1 - \sqrt{-1}(\varepsilon^{-\eta}-1)\right]^{k-i-1} - \left[1 + \sqrt{-1}(\varepsilon^{-\eta}-1)\right]^{k-i-1} \right\} d\eta -$$

$$- 2\frac{\Theta(i-1-k)}{2\sqrt{2}\sqrt{-1}} \int_0^{\sqrt{2}} \frac{\varepsilon^{\eta(j-q)}}{\varepsilon^{-\eta}-1} \times$$

$$\left\{ \left[1 - \sqrt{-1}(\varepsilon^{-\eta}-1)\right]^{k-i} - \left[1 + \sqrt{-1}(\varepsilon^{-\eta}-1)\right]^{k-i} \right\} d\eta -$$

$$- 2\frac{\Theta(i-k)}{2\sqrt{2}\sqrt{-1}} \int_0^{\sqrt{2}} \frac{\varepsilon^{\eta(j-1-q)}}{\varepsilon^{-\eta}-1} \times$$

$$\times \left\{ \left[1 - \sqrt{-1}(\varepsilon^{-\eta}-1)\right]^{k-i-1} - \left[1 + \sqrt{-1}(\varepsilon^{-\eta}-1)\right]^{k-i-1} \right\} d\eta +$$

$$+ 2\frac{\Theta(i-k)}{2\sqrt{2}\sqrt{-1}} \int_0^{\sqrt{2}} \frac{\varepsilon^{\eta(j-q)}}{\varepsilon^{-\eta}-1} \times$$

$$\times \left\{ \left[1 - \sqrt{-1}(\varepsilon^{-\eta}-1)\right]^{k-i-1} - \left[1 + \sqrt{-1}(\varepsilon^{-\eta}-1)\right]^{k-i-1} \right\} d\eta =$$

$$= \frac{\Theta(i-2-k) - \Theta(i-k-1)}{2\sqrt{2}\sqrt{-1}} \int_0^{\sqrt{2}} \frac{\varepsilon^{\eta(j-q)}}{\varepsilon^{-\eta}-1} \times$$

$$\times \left\{ \left[1 - \sqrt{-1}(\varepsilon^{-\eta}-1)\right]^{k-i+1} - \left[1 + \sqrt{-1}(\varepsilon^{-\eta}-1)\right]^{k-i+1} \right\} d\eta +$$

$$+ \frac{\Theta(i-1-k) - \Theta(i-k)}{2\sqrt{2}\sqrt{-1}} \int_0^{\sqrt{2}} \frac{\varepsilon^{\eta(j-q)}}{\varepsilon^{-\eta}-1} \times$$

$$\times \left\{ \left[1 - \sqrt{-1}(\varepsilon^{-\eta}-1)\right]^{k-i+1} - \left[1 + \sqrt{-1}(\varepsilon^{-\eta}-1)\right]^{k-i+1} \right\} d\eta -$$

$$- 2\frac{\Theta(i-k-1) - \Theta(i-k)}{2\sqrt{2}\sqrt{-1}} \int_0^{\sqrt{2}} \frac{\varepsilon^{\eta(j-q)}}{\varepsilon^{-\eta}-1} \times$$

$$\times \left\{ \left[1 - \sqrt{-1}(\varepsilon^{-\eta}-1)\right]^{k-i} - \left[1 + \sqrt{-1}(\varepsilon^{-\eta}-1)\right]^{k-i} \right\} d\eta +$$

$$+ \frac{\Theta(i-k)}{2\sqrt{2}\sqrt{-1}} \int_0^{\sqrt{2}} \frac{\varepsilon^{\eta(j-q)}}{\varepsilon^{-\eta}-1} \times$$

$$\times \left\{ \left[1 - \sqrt{-1}(\varepsilon^{-\eta} - 1)\right]^{k-i+1} - \left[1 + \sqrt{-1}(\varepsilon^{-\eta} - 1)\right]^{k-i+1} + \right.$$

$$+ \varepsilon^{-2\eta} \left[1 - \sqrt{-1}(\varepsilon^{-\eta} - 1)\right]^{k-i-1} - \varepsilon^{-2\eta} \left[1 + \sqrt{-1}(\varepsilon^{-\eta} - 1)\right]^{k-i-1} -$$

$$- 2 \left[1 - \sqrt{-1}(\varepsilon^{-\eta} - 1)\right]^{k-i} + 2 \left[1 + \sqrt{-1}(\varepsilon^{-\eta} - 1)\right]^{k-i} -$$

$$- 2\varepsilon^{-\eta} \left[1 - \sqrt{-1}(\varepsilon^{-\eta} - 1)\right]^{k-i-1} + 2\varepsilon^{-\eta} \left[1 + \sqrt{-1}(\varepsilon^{-\eta} - 1)\right]^{k-i-1} +$$

$$+ 2 \left[1 - \sqrt{-1}(\varepsilon^{-\eta} - 1)\right]^{k-i-1} - 2 \left[1 + \sqrt{-1}(\varepsilon^{-\eta} - 1)\right]^{k-i-1} \right\} .$$

$$(A.42)$$

Since the expression inside the last bracket is the characteristic equation, it is equal to zero. On the other hand, according to the fact

$$\Theta(i - k - 2) - \Theta(i - k - 1) = \begin{cases} 0 & i \geq k + 2 \\ -1 & i = k + 1 \\ 0 & i \leq k \end{cases}$$

the first term is equal to zero. Similarly because of

$$\Theta(i - k - 1) - \Theta(i - k) = \begin{cases} 0 & i \geq k + 1 \\ -1 & i = k \\ 0 & i \leq k - 1 \end{cases}$$

the third term is also equal to zero.

Finally the second term is

$$\frac{\Theta(i - k - 1) - \Theta(i - k)}{2\sqrt{2}\sqrt{-1}} \int_0^{\sqrt{2}} \frac{\varepsilon^{\eta(j-q)}}{\varepsilon^{-\eta} - 1} \times$$

$$\times \left\{ \left[1 - \sqrt{-1}(\varepsilon^{-\eta} - 1)\right]^{k-i+1} - \left[1 + \sqrt{-1}(\varepsilon^{-\eta} - 1)\right]^{k-i+1} \right\} d\eta =$$

$$= -\frac{\delta(i - k)}{2\sqrt{2}\sqrt{-1}} \int_0^{\sqrt{2}} \frac{\varepsilon^{\eta(j-q)}}{\varepsilon^{-\eta} - 1} \times$$

$$\left\{ \left[1 - \sqrt{-1}(\varepsilon^{-\eta} - 1)\right]^{k-i+1} - \left[1 + \sqrt{-1}(\varepsilon^{-\eta} - 1)\right]^{k-i+1} \right\} d\eta =$$

$$= -\frac{\delta(i - k)}{2\sqrt{2}\sqrt{-1}} \int_0^{\sqrt{2}} \frac{\varepsilon^{\eta(j-q)}}{\varepsilon^{-\eta} - 1} \left\{ -2\sqrt{-1}(\varepsilon^{-\eta} - 1) \right\} d\eta =$$

$$= \delta(i - k)\frac{1}{\sqrt{2}} \int_0^{\sqrt{2}} \varepsilon^{\eta(j-q)} d\eta = \delta(i - k)\delta(j - q) \quad (A.43)$$

where

$$\delta(i - k) = \begin{cases} 1 & i = k \\ 0 & i \neq k, . \end{cases} \qquad (A.44)$$

From (A.42) and (A.43) we see that

$$\Delta^* V_{i-k,j-q} \equiv V_{i-k-2,j-q} + V_{i-k,j-q-2} - 2V_{i-k-1,j-q} -$$
$$- 2V_{i-k,j-q-1} + 2V_{i-k,j-q} = \delta(i - k)\delta(j - q) . \quad (A.45)$$

Therefore from (A.41) we get the following

$$U_{kq} = 2\sum_{j=0}^{m} U_{n+1,j} V_{n-k,j-q} + 2\sum_{i=0}^{n} \left(U_{i(m+1)} V_{i-k,m-j} - U_{i0} V_{i-k,-1-q} \right) -$$
$$- \sum_{j=0}^{m} \left(U_{n+1,j} V_{n-k-1,j-q} + U_{n+2,j} V_{n-k,j-q} \right) -$$
$$- \sum_{i=0}^{n} \left(U_{i(m+1)} V_{i-k,m-1-q} + U_{i(m+2)} V_{i-k,m-q} -$$
$$- U_{i0} V_{i-k,-2-q} - U_{i1} V_{i-k,-1-q} \right) \quad k = 0, 1, \dots n \quad q = 0, 1, \dots, m .$$
$$(A.46)$$

The last expression i.e., (A.46) implies the following theorem.

Theorem. *Any definite solution of the equation (A.29) in the discrete polygon $i = 0, 1, \dots, m$, $j = 0, 1, \dots, m$ can be found via the expression (A.46).*

Remark. If $j - q < 0$ then by the change of variable $\varepsilon^{-\eta} - 1 = \xi$ in the fundamental solution (A.39) we get the following

$$V_{i-k,j-q} = \frac{\Theta(i - k)}{2\sqrt{2}\sqrt{-1}} \int_c \frac{(1 + \xi)^{q-j-1}}{-2\pi\sqrt{-1}\xi} \times$$
$$\times \left\{ \left[1 - \sqrt{-1}(\varepsilon^{-\eta} - 1) \right]^{k-i-1} - \left[1 + \sqrt{-1}(\varepsilon^{-\eta} - 1) \right]^{k-i-1} \right\} d\xi$$

where c is a unit circle centered at $(-1, 0)$ and integration has counterclockwise direction,

On the other hand by considering

$$\frac{(1 - \sqrt{-1}\xi)^{k-i-1} - (1 + \sqrt{-1}\xi)^{k-i-1}}{\xi} =$$

$$= \frac{(1 + \sqrt{-1}\xi)^{1+i-k} - (1 - \sqrt{-1}\xi)^{1+i-k}}{\xi(1 + \xi^2)^{1+i-k}}$$

and the fact that the expression $(1 + \xi)^{q-i-1}$ with $q - i - 1 \geq 0$ is analytic on circle c and its inside, the above integral is equal to zero, that is

$$V_{i-k, j-q} = 0 .$$

In fact

$$V_{i-k, j-q} = 0 \quad \text{if } i < k \text{ or } j < q . \tag{A.47}$$

Now we add the following boundary conditions to the equation (A.29)

$$U_{n+1,q} - U_{0q} = \alpha_q$$

$$U_{n+2,q} - U_{1q} = \beta_q \qquad q = 0, 1, \ldots, m$$

$$U_{k,m+1} - U_{k0} = \ell_k$$

$$U_{k,m+2} - U_{k1} = t_k \qquad k = 0, 1, \ldots, n .$$

Then by (A.47) the general solution of (A.46) becomes

$$U_{kq} = 2 \sum_{j=0}^{m} U_{n+1,j} V_{n-k,j-q} + 2 \sum_{i=0}^{n} U_{i,m+1} V_{i-k,m-j} -$$

$$- \sum_{j=0}^{m} \left(U_{n+1,j} V_{n-k-1,j-q} + U_{n+2,j} V_{n-k,j-q} \right) -$$

$$- \sum_{i=0}^{n} \left(U_{i(m+1)} V_{i-k,m-1-q} + U_{i(m+2)} V_{i-k,m-q} \right) .$$

If we consider here the boundary conditions, we get

$$
\begin{aligned}
U_{kq} = {} & 2\sum_{j=0}^{m} \left(U_{0j} + \alpha_j\right) V_{n-k,j-q} + 2\sum_{j=0}^{n} \left(U_{i0} + \ell_i\right) V_{i-k,m-j} - \\[2mm]
& - \sum_{j=0}^{m} \left[\left(U_{0j} - \alpha_j\right) V_{n-k-1,j-q} + \left(U_{1j} + \beta_j\right) V_{n-k,j-q}\right] - \\[2mm]
& - \sum_{i=0}^{n} \left[\left(U_{i0} + \ell_i\right) V_{i-k,m-1-q} + \left(U_{i1} + t_i\right) V_{i-k,m-q}\right] \\[2mm]
& k = 0, 1, \ldots, n \qquad q = 0, 1, \ldots, m \ .
\end{aligned}
\tag{A.48}
$$

In this last expression, if we first take $k = 0$, $k = 1$, then $q = 0$, $q = 1$ and then from the resulting system, determine the values of U_{0j}, U_{1j}, U_{i0}, and U_{i1}, then we obtain the solution of (A.29) and (A.48).

A.7   Integral over an arbitrary broken curve in discrete plane Z^2

Consider the function f on $Z \times Z$. We wish to evaluate its integral over a broken polygon. To this end we first need the following definition and notation.

Definition and notation. Suppose that $n+1$ points $A_0, A_1, \ldots, A_n$ in the discrete plane Z^2 are given such that the distance between two consecutive points A_{k-1} and A_k $(1 \leq k \leq n)$ is equal to unity, then the set of points is called a broken curve in Z^2, and is denoted by Γ. If in addition, the points A_0 and A_n are chosen to be the initial and final points respectively then Γ is called a directed curve, in this case we denote it by $\vec{\Gamma}$. It is clear from this definition that the directed length of $\overrightarrow{A_{k-1}A_k}$ is one of the four cases $+1, -1, +i, -i$. We denote this length by $\Delta\ell_k$ (see the following figure)

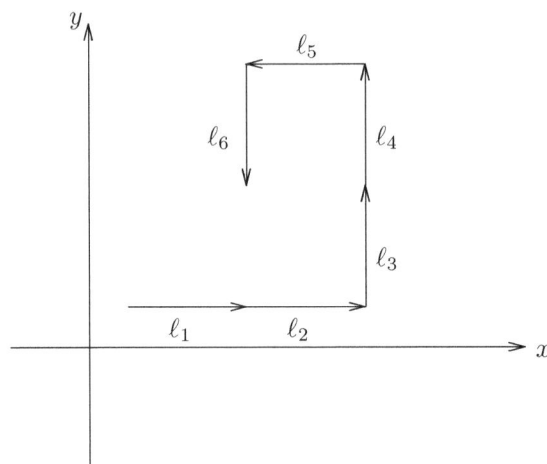

For example, in the above figure we have

$$\Delta\ell_1 = 1 \,, \quad \Delta\ell_3 = i \quad \Delta\ell_5 = -1 \,, \quad \Delta\ell_6 = -i \,.$$

Now suppose that the point A_k ($0 \le k \le n$) has the coordinate (x_k, y_k) in Z^2. Then we have the following definition.

Definition.   Suppose that A_k ($k = 0, 1, \ldots, n$) is the set of the points on the directed broken curve $\vec{\Gamma}$ in Z^2 where the complex function is defined, then we define the finite sum

$$\sum_{k=0}^{n-1} f(x_k + iy_k)\Delta\ell_k$$

as the integral of f over $\vec{\Gamma}$ and we denote it by

$$\int_{\Gamma} f(x + iy)\Delta\ell \,.$$

A.8   Discrete integral of the second type over a broken curve in Z^2

Unlike the integral defined on previous section, in this section we are going to introduce another type of integral which depends not only on the value of

$\Delta \ell_k$ but its direction as well. (These directions are the positive and negative directions of the coordinate axes.) To define this integral we first need some pre-definitions.

Definition 1. If $x \in Z$ is an arbitrary point, then we call the line segments $[x, x+1]$ and $[x+1, x]$ as open curves and we denote them by $c(x)$ and $\bar{c}(x)$. Besides if we consider these two segments simultaneously, then the resulting system is called a simple closed curve and is denoted by $\Gamma(x)$.

Definition 2. If f is a function with domain Z and $\Gamma(x)$ is a simple closed curve in Z, then we define the integral of f over $\Gamma(x)$ as

$$\int_{\Gamma(x)} f(\xi)\Delta\xi = f(x) \cdot 1 + f(x)(-1) \ .$$

In words, to evaluate the integral we take the value of f at initial point of the open interval in the positive direction of Z–axis plus the value of f at final point in the negative direction of the same axis.

Similarly we may define simple open and closed intervals in Z^2.

Definition 3. Let $(x, y) \in Z^2$ be any point, then the line segments obtained by the consecutive points of $[(x, y), (x+1, y), (x+1, y+1)]$ and $[(x, y), (x, y+1), (x+1, y+1)]$ are called the simple open curves in the positive direction. The simple open curves are similarly defined.

Definition 4. Let $(x, y) \in Z^2$ be any point, then the five consecutive points $[(x, y), (x+1, y), (x+1, y+1), (x, y+1), (x, y)]$ with the initial and final points (x, y) is called a simple closed curve in Z^2, and denoted by $\Gamma(x, y)$ (see the following figure).

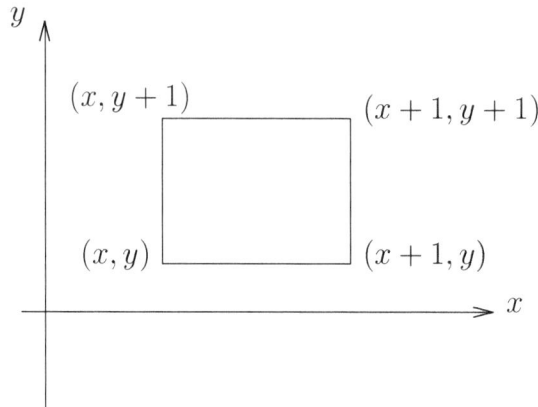

Definition 5. Let f be a function defined on Z^2 and let $\Gamma(x, y)$ is a simple closed curve in Z^2, then the integral of f over $\Gamma(x, y)$ is defined by

$$\int_{\Gamma(x,y)} f(\xi, \eta)\Delta\Gamma = f(x, y) \cdot 1 + f(x + 1, y)i + f(x, y + 1)(-1) + f(x, y)(-i) .$$

$$(A.49)$$

It is clear from this definition that if the curve of integration is the form of

$$\Gamma(x, y) = [(x, y), (x + 1, y)] \cup [(x + 1, y), (x, y)] .$$

then the value of the integral is equal to zero, i.e.,

$$\int_{\Gamma(x,y)} f(u, v)\Delta\Gamma = f(x, y) \cdot 1 + f(x, y) \cdot (-1) = 0 .$$

This conclusion cannot be true for any function. So to determine the functions satisfying the above property is of special importance.

In order to do this, consider the function defined by

$$f(x, y) = U(x, y) + iV(x, y) .$$

$$(A.50)$$

By integrating this function over the simple closed curve $\Gamma(x, y)$, we have

$$\int_{\Gamma(x,y)} f(\xi, \eta)\Delta\Gamma = U(x, y) + iV(x, y) + i\left[U(x+1, y) + iV(x+1, y)\right] -$$

$$- \left[U(x, y+1) + iV(x, y+1)\right] - i\left[U(x, y) + iV(x, y)\right]$$

$$= \{U(x, y) - U(x, y+1) - V(x+1, y) + V(x, y)\} +$$

$$+ i\{V(x, y) - V(x, y+1) + U(x+1, y) - U(x, y)\}$$

$$= \left(\frac{-\partial U}{\partial y} - \frac{\partial V}{\partial x}\right) + i\left(\frac{-\partial V}{\partial y} + \frac{\partial U}{\partial x}\right) .$$

$$(A.51)$$

The following theorem is an immediate consequence of the above equality.

Theorem. *The necessary and sufficient condition for the integral of the complex discrete function $f = u + iv$ over the simple closed curve $\Gamma(x, y)$ is equal to zero, is that the functions U and V satisfy in the discrete Cauchy-Riemann equations.*

Proof. Let

$$\begin{cases} \frac{\partial U}{\partial x} = \frac{\partial V}{\partial y} \\ \frac{\partial U}{\partial y} = -\frac{\partial V}{\partial x} . \end{cases} \tag{A.52}$$

Then from (A.51) it is clear that

$$\int_{\Gamma(x,y)} f(\xi, \eta)\Delta\Gamma = 0 . \tag{A.53}$$

Conversely if (A.53) holds, then the equation of (A.52) is immediate. This theorem can be easily extended over any broken closed curve in Z^2. In fact we have the following theorem.

Theorem. *Let the function $f = u + iv$ be analytic on a bounded region containing the broken closed curve c. Then the integral of f over c is equal to zero.*

Proof. By induction on the number of the simple closed curves in c we observe that if $\Gamma(x, y)$ is a simple closed curve in c, then by the previous theorem the integral of f over $\Gamma(x, y)$ is equal to zero. Now suppose that the integral of f over the broken closed curve $c' = c \setminus \Gamma(x, y)$ is equal to zero. By the additive property of the integral with respect to the curves of integration we have

$$\int_c f(\xi, \eta)\Delta\Gamma = \int_{\Gamma(x,y)} f(\xi, \eta)\Delta\Gamma + \int_{c'} f(\xi, \eta)\Delta\Gamma = 0 + 0$$

where the first integral by the previous theorem and the second integral by the induction hypothesis are equal to zero. This theorem can be regarded as the discrete analog of the Cauchy theorem in complex analysis.

A.9   The cubic root of a complex number

In Section 4 of Chapter 1 the calculation of the second roots of a complex number was presented. In this article we wish to examine the third roots of a complex number in algebraic form. To this end, let $a + ib$ be an arbitrary complex number. If $x + iy$ is a third root of $a + ib$, then

$$\sqrt[3]{a + ib} = x + iy \ . \tag{A.54}$$

By cubing both sides of this equality we get the following complex equation

$$x^3 - 3xy^2 + 3x^2yi - y^3i = a + ib$$

which is in turn equivalent to the non-linear system of equations

$$\begin{cases} x^3 - 3xy^2 = a \\ 3x^2y - y^3 = b \ . \end{cases} \tag{A.55}$$

If we find x from the second equation and substitute the result in the first one, we have

$$\pm\frac{y^3 + b}{3y}\sqrt{\frac{y^3 + b}{3y}} \pm 3y^2\sqrt{\frac{y^3 + b}{3y}} = a \ .$$

By canceling the radicals, we obtain

$$\frac{(b - 8y^3)^2(y^3 + b)}{27y^3} = a^2$$

or equivalently

$$64(y^3)^3 + 48b(y^3)^2 - (15b^2 + 27a^2)y^3 + b^3 = 0 \ .$$

By taking $y^3 = z - \frac{b}{4}$ we have

$$64\left(z^3 - \frac{36}{4}z^2 + \frac{3b^2}{16}z - \frac{b^3}{64}\right) + 48b\left(z^2 - \frac{b}{2}z + \frac{b^2}{16}\right) -$$
$$- (15b^2 + 27a^2)\left(z - \frac{b}{4}\right) + b^3 = 0 \qquad .$$

Rearranging the terms of the left-hand side of this equation in terms of decreasing powers of z and simplifying we get

$$z^3 - \left(\frac{3}{4}\right)^3(a^2 + b^2)z + \frac{3^3}{4^4}b(a^2 + b^2) = 0 \ . \qquad (A.56)$$

Now by Cardan's formula, that is

$$z = \sqrt[3]{\frac{-q}{2} + \sqrt{\frac{q^2}{4} + \frac{p^3}{27}}} + \sqrt[3]{\frac{-q}{2} - \sqrt{\frac{q^2}{4} + \frac{p^3}{27}}} \ . \qquad (A.57)$$

We can obtain the following solution for (A.56)

$$z = \sqrt[3]{-\frac{3^3}{2 \cdot 4^4}b(a^2 + b^2) + \frac{3^3}{4^4 \cdot 2}i(a^2 + b^2)a+}$$
$$+ \sqrt[3]{-\frac{3^3}{2 \cdot 4^4}b(a^2 + b^2) - \frac{3^3}{4^4 \cdot 2}i(a^2 + b^2)a} \qquad (A.58)$$
$$= -\frac{3}{8}\sqrt[3]{a^2 + b^2}\left(\sqrt[3]{(b + ai)} + \sqrt[3]{b - ai}\right) \ .$$

As we observe, to determine the third root of a complex number gives rise to a solution of a cubic equation and conversely the solution of a cubic equation reduces to the determining of the third root of a complex number. This means that finding the real root of a cubic equation by Cardan's formula is impossible. (In fact this real root appears as the sum of the third roots of two conjugate complex number.) Therefore to determine the third root of a complex number remains unsolved.

A.10   The spectral decomposition of a matrix by projection operators

Let A be a $n \times n$ matrix with real entries, then the polynomial

$$\det(A - \lambda I) \tag{A.59}$$

is called the characteristic polynomial of A and the equation

$$\det(A - \lambda I) = 0 \tag{A.60}$$

is called its characteristic equations, where I is the $n \times n$ identity matrix. Let A be a symmetric matrix, i.e., if $A = [A_{ij}]$ then

$$A_{ij} = A_{ji} \qquad \forall\, i, j \ .$$

Also suppose that the eigenvalues of A i.e., the roots of the characteristic equation are all distinct and different from zero, that is

$$\lambda_p \neq \lambda_q \text{ for } p \neq q \text{ and } \lambda_i \neq 0, \qquad i = 1, \ldots, n$$

(recall that these roots are all real, since A is symmetric). Then the homogeneous systems of equations

$$A x_k = \lambda_k x_k \qquad k = 1, \ldots, n \tag{A.61}$$

all have non-trivial solutions

$$x_k = \left(x_k^1, \ldots, x_k^n\right)^T \qquad k = 1, \ldots, n \tag{A.62}$$

where T stands for the transpose of a matrix. This solution is called the corresponding eigenvector of the eigenvalue of λ. Instead of the system (A.61) one pick up cofactors of certain row of the matrix

$$A - \lambda_k I = (A_{ij} - \lambda_k \delta_{ij}) \ .$$

Since the matrix A is symmetric, we can make these solutions into an orthonormal set, i.e., for each p, q we can write

$$x_p^T x_q = \delta_{pq} \ . \tag{A.63}$$

If instead of the above product we use

$$x_p \cdot x_q^T \qquad p, q = 1, \ldots, n \ . \tag{A.64}$$

Then for $n \times n$ matrices

$$p_k = x_k \cdot x_k^T \qquad k = 1, \ldots, n \ . \tag{A.65}$$

We have the following identities

a) $p_k^2 = p_k$

b) $p_k p_m = 0$ if $k \neq m$

c) $I = \sum\limits_{k=1}^{n} p_k$

d) $A = \sum\limits_{k=1}^{n} \lambda_k p_k$

Proof. In fact,

$$
\begin{aligned}
p_k p_m &= \left(x_k \cdot x_k^T \right) \left(x_m \cdot x_m^T \right) \\
&= x_k \left(x_k^T \cdot x_m \right) x_m^T \\
&= x_k \delta_{km} x_m^T =
\begin{cases}
p_k & k = m \\
0 & k \neq m \ .
\end{cases}
\end{aligned}
\tag{A.66}
$$

On the other hand, since the vectors in (A.62) are linearly independent and their number is equal to n, each arbitrary vector can be expressed as a linear combination of these vectors, i.e., we have

$$x = \sum_{k=1}^{n} c_k x_k .$$ (A.67)

To determine the coefficients c_k, we multiply both sides of this equation by x_q^T from the left, so

$$
\begin{aligned}
x_q^T \cdot x &= \sum_{k=1}^{n} x_q^T (c_k x_k) = \sum_{k=1}^{n} c_k (x_q^T x_k) \\
&= \sum_{k=1}^{n} c_k \delta_{qk} = c_q .
\end{aligned}
$$ (A.68)

By substituting the values of c_k from (A.68) into (A.67) we get

$$x = \sum_{k=1}^{n} (x_k^T \cdot x) x_k .$$ (A.69)

We can rewrite this expression as

$$x = \sum_{k=1}^{n} x_k (x_k^T \cdot x)$$

which in turn can be expressed by using the associative property of matrices as

$$x = \sum_{k=1}^{n} (x_k \cdot x_k^T) x .$$

Since x is an arbitrary vector, by (A.65) we have

$$I = \sum_{k=1}^{n} p_k .$$ (A.70)

Finally to prove the last assertion we observe that if we multiply both sides of (A.61) by x_k^T from the right and sum up the resulting equations we get

$$\sum_{k=1}^{n} A x_k \cdot x_k^T = \sum_{k=1}^{n} \lambda_k x_k x_k^T .$$

Again by (A.65) we have

$$A \sum_{k=1}^{n} p_k = \sum_{k=1}^{n} \lambda_k p_k$$

which by using (A.70) gives rise to

$$A = \sum_{k=1}^{n} \lambda_k p_k \ . \tag{A.71}$$

In this way, we determined the spectral decomposition of a symmetric matrix.

We are going now to utilize this decomposition for the solution of the system of equation

$$Ax = b \tag{A.72}$$

where $A = [A_{ij}]$ is an $n \times n$ real symmetric matrix and $b = (b_1, \ldots, b_n)^T$ is an $n \times 1$ column matrix. To this end, if we substitute (A.67) and (A.71) into (A.72) and use the definition (A.65) then

$$\sum_{k=1}^{n} \lambda_k x_k x_k^T \sum_{p=1}^{n} c_p x_p = b \ .$$

By commuting the two sigma , we get

$$\sum_{p=1}^{n} \left(\sum_{k=1}^{n} \lambda_k x_k x_k^T x_p \right) c_p = b \ .$$

Now by (A.63) we have

$$\sum_{p=1}^{n} \left(\sum_{k=1}^{n} \lambda_k x_k \delta_{kp} \right) c_p = b$$

or equivalently

$$\sum_{p=1}^{n} \lambda_p x_p c_p = b \ .$$

Finally if we multiply the above equality by x_k^T from the left we get

$$x_k^T b = \sum_{p=1}^{n} \lambda_p x_k^T x_p c_p = \sum_{p=1}^{n} \lambda_p \delta_{kp} c_p = \lambda_k c_k \ . \tag{A.73}$$

Since

$$\lambda_k \neq 0 \qquad k = 1, \ldots, n$$

from (A.73) we get

$$c_k = \frac{1}{\lambda_k} x_k^T b \ . \tag{A.74}$$

By substituting this value in (A.67) for the solution we have

$$x = \sum_{k=1}^{n} \frac{1}{\lambda_k} x_k \cdot x_k^T b \tag{A.75}$$

or

$$x = \sum_{k=1}^{n} \frac{1}{\lambda_k} p_k b \tag{A.76}$$

which is a concise expression.

Now let for one k, we have $\lambda_k = 0$. Without loss of generality we assume that $\lambda_n = 0$, then by (A.73) we have

$$x_n^T b = 0 \ . \tag{A.77}$$

Note that if (A.77) holds, then the system (A.72) always has infinitely many solutions, in fact, in the case of existence, the solution is in the form

$$x = \sum_{k=1}^{n-1} \frac{1}{\lambda_k} p_k b + x_n c_n \tag{A.78}$$

where c_n is an arbitrary constant. To derive this expression, we first observe that

$$
\begin{aligned}
Ax &= \sum_{k=1}^{n-1} \lambda_k p_k \sum_{m=1}^{n-1} \frac{1}{\lambda_m} p_m b \\
&= \sum_{k=1}^{n-1} \sum_{m=1}^{n-1} \frac{\lambda_k}{\lambda_m} p_k p_m b \\
&= \sum_{k=1}^{n-1} p_k b = \left(\sum_{k=1}^{n} p_k - p_n \right) b \\
&= (I - p_n) b = b - p_n b .
\end{aligned}
\tag{A.79}
$$

On the other hand, if we multiply the equality (A.77) by x_n from the left, we have

$$
(x_n \cdot x_n^T) b = p_n b = 0 .
\tag{A.80}
$$

Now the result follows from (A.79) and (A.80).

We conclude this section by constructing the spectral decomposition of a $n \times n$ real matrix with distinct and non-zero eigenvalues. It can be easily shown that for any $n \times n$ real matrix A both matrices A and A^T have the same eigenvalues, that is

$$
\det(A - \lambda I) = \det(A^T - \lambda I) .
$$

If we denote the eigenvalues of A by λ_k $(k = 1, \ldots, n)$ and the eigenvectors of A and A^T by x_k $(k = 1, \ldots, n)$ and y_k $(k = 1, \ldots, n)$ respectively, then

$$
\begin{aligned}
Ax_k &= \lambda_k x_k \\
A^T y_k &= \lambda_k y_k
\end{aligned}
\qquad (k = 1, 2, \ldots, n) .
\tag{A.81}
$$

We orthonormalize these vectors such that

$$
y_q^T x_p = \delta_{pq} .
\tag{A.82}
$$

Then according to the definition

$$p_k = x_k y_k^T \qquad k = 1, \dots, n \tag{A.83}$$

we have the following

$$p_k^2 = p_k$$

$$p_m p_k = 0 , \qquad m \neq k$$

$$I = \sum_{k=1}^{n} p_k$$

$$A = \sum_{k=1}^{n} \lambda_k p_k .$$

Proof. In fact,

$$p_m p_k = x_m y_m^T x_k y_k^T = x_m \delta_{mk} y_k^T$$

$$= \begin{cases} p_k & m = k \\ 0 & m \neq k . \end{cases} \tag{A.84}$$

Now for the proof of the third equality, we multiply both sides of

$$x = \sum_{k=1}^{n} x_k c_k \tag{A.85}$$

by y_p^T from the right we have

$$y_p^T x = \sum_{k=1}^{n} y_p^T x_k c_k = \sum_{k=1}^{n} \delta_{pk} c_k = c_p .$$

By substituting the value of c_p in (A.85) we get

$$x = \sum_{k=1}^{n} x_k y_k^T x = \sum_{k=1}^{n} p_k x .$$

Since x is an arbitrary vector, we get

$$\sum_{k=1}^{n} p_k = I \ .$$

(A.86)

Finally for the proof of (IV) if we multiply the first expression of (A.81) by y_k^T from the right and sum up the terms, we get

$$\sum_{k=1}^{n} A x_k y_k^T = \sum_{k=1}^{n} \lambda_k x_k y_k^T$$

or equivalently

$$A \sum_{k=1}^{n} p_k = \sum_{k=1}^{n} \lambda_k p_k$$

which by using (A.86) gives rise to

$$A = \sum_{k=1}^{n} \lambda_k p_k$$

(A.87)

and this is what we had to prove.

Bibliography

[1] Chalice, D. R., How to differentiate and integrate sequences, The American Mathematical Monthly, Vol. 108 Number 10, December 2001.

[2] Blanchard, P., Devaney R. L., Hall, G. R., Differential equations, Brooks-Cole, Boston, 1998.

[3] Kelly, W. G., Peterson, A. C., Difference equations: An introduction with applications, Academic press, 1991.

[4] Vorob'ev, N. N, Fibonacci numbers, English edition, New York, Blaisdell publication 1991.

[5] Schoenberg, I. J., Mathematical time exposures, The Mathematical Association of America, 1982.

[6] Sawyer, W. W., Prelude to mathematics, Dover publication, 1982.

[7] Vladimirov, V. S, Mathematical physics equations, Nauka, Moscow 1981 (In Russian).

[8] Liu, C. L., Elements of discrete mathematics, McGraw-Hill, New York, 1977.

[9] Riordan, J., Combinatorial identities, JOHN WILEY SONS, New York-London-Sydney, 1968.

[10] Levy, H., Lessman. F., Finite difference equations, Dover publication, New Yourk, 1961.

Index

www.ingramcontent.com/pod-product-compliance
Lightning Source LLC
Chambersburg PA
CBHW041711210326
41598CB00007B/617